RESERVOIR SEDIMENTATION

SPECIAL SESSION ON RESERVOIR SEDIMENTATION OF THE SEVENTH INTERNATIONAL CONFERENCE ON FLUVIAL HYDRAULICS (RIVER FLOW 2014), IAHR COMMITTEE ON FLUVIAL HYDRAULICS, EPFL LAUSANNE, SWITZERLAND, 3–5 SEPTEMBER 2014

Reservoir Sedimentation

Editors

Anton J. Schleiss, Giovanni De Cesare, Mário J. Franca & Michael Pfister

Laboratoire de Constructions Hydrauliques (LCH)
École Polytechnique Fédérale de Lausanne (EPFL), Switzerland

CRC Press
Taylor & Francis Group
Boca Raton   London   New York   Leiden

CRC Press is an imprint of the
Taylor & Francis Group, an **informa** business

A BALKEMA BOOK

Cover photo: Moiry Dam near Zinal (VS), Switzerland. Courtesy: Michael Pfister

CRC Press/Balkema is an imprint of the Taylor & Francis Group, an informa business

© 2014 Taylor & Francis Group, London, UK

Typeset by V Publishing Solutions Pvt Ltd., Chennai, India
Printed and bound in Great Britain by CPI Group (UK) Ltd, Croydon, CR0 4YY

Published by: CRC Press/Balkema
               P.O. Box 11320, 2301 EH Leiden, The Netherlands
               e-mail: Pub.NL@taylorandfrancis.com
               www.crcpress.com – www.taylorandfrancis.com

ISBN: 978-1-138-02675-9 (Hardback + CD-ROM)
ISBN: 978-1-315-73691-4 (eBook PDF)

Table of contents

Reservoir Sedimentation – Schleiss et al. (Eds)
© *2014 Taylor & Francis Group, London, ISBN 978-1-138-02675-9*

Preface

Despite the mechanisms of reservoir sedimentation being well known for a long time, sustainable and preventive measures are rarely taken into consideration in the design of new reservoirs. To avoid operational problems of powerhouses, sedimentation is often treated for existing reservoirs with measures which are efficient only for a limited time. Since most of the measures will lose their effect, the sustainable operation of the reservoir and thus the water supply, as well as production of valuable peak energy is endangered. Today's worldwide yearly mean loss of reservoir storage capacity due to sedimentation is already higher than the increase of capacity by the addition of new reservoirs for irrigation, drinking water, and hydropower. Depending on the region, it is commonly accepted that about 1–2% of the worldwide capacity is lost annually. In Asia, for example, 80% of the useful storage capacity for hydropower production will be lost by 2035. The main sedimentation process in narrow and long reservoirs is the formation of turbidity currents, transporting fine sediments near to the dam every flood season, increasing sediment levels up to 1 m per year. The outlet devices including intakes and bottom outlets are often affected after only 40 to 50 years of operation, even in catchments with moderate surface erosion rates. The effects of climate change will considerably increase the sediment yield of reservoirs in the future, mainly in Alpine regions due to glacier retreat and melting of permafrost grounds.

The today's challenge of dam owners and engineers is to guarantee with adequate mitigation measures, the future sustainable use of the vital reservoirs supplying water for drinking, food and energy production. Research and development is still urgently needed to identify efficient mitigation measures adapted to the main sedimentation processes involved in reservoirs.

During the Seventh International Conference on Fluvial Hydraulics "River Flow 2014" at École Polytechnique Fédérale de Lausanne (EPFL), Switzerland, scientists and professionals from all over the world addressed the challenge of reservoir sedimentation in a special session and exchanged their knowledge and experiences. The conference was organized under the auspices of the Committee on Fluvial Hydraulics of the International Association for Hydro-Environment Engineering and Research (IAHR). Invited and selected contributions, which give an overview on the latest developments and research regarding reservoir sedimentation including case studies, are presented in this book, hoping that they can contribute to better sustainable use of the vital reservoirs worldwide. We acknowledge the support of the Swiss Federal Office for the Environment, BG Consulting Engineers, and Hydro Exploitation SA as main sponsors for the Proceedings and the River Flow 2014 Conference. Further sponsoring was obtained from e-dric.ch, IM & IUB Engineering, Basler & Hofmann, Aqua-Vision Engineering and Met-Flow SA.

The Laboratory of Hydraulic Constructions (LCH) of EPFL carried out the organization of the special session on reservoir sedimentation in the frame of River Flow 2014.

Prof. Dr. Anton J. Schleiss, *Conference Chairman*
Dr. Giovanni De Cesare, *Co-Chairman*
Dr. Mário J. Franca, *Co-Chairman*
Dr. Michael Pfister, *Co-Chairman*

Reservoir Sedimentation – Schleiss et al. (Eds)
© *2014 Taylor & Francis Group, London, ISBN 978-1-138-02675-9*

Organization

MEMBERS OF THE INTERNATIONAL SCIENTIFIC COMMITTEE

Walter H. Graf, Switzerland *(Honorary Chair)*

Jorge D. Abad, *USA*
Jochen Aberle, *Norway*
Claudia Adduce, *Italy*
Mustafa Altınakar, *USA*
Christophe Ancey, *Switzerland*
William K. Annable, *Canada*
Aronne Armanini, *Italy*
Francesco Ballio, *Italy*
Roger Bettess, *UK*
Koen Blanckaert, *China*
Robert Michael Boes, *Switzerland*
Didier Bousmar, *Belgium*
Benoît Camenen, *France*
António Heleno Cardoso, *Portugal*
Hubert Chanson, *Australia*
Qiuwen Chen, *China*
Yee-Meng Chiew, *Singapore*
George S. Constantinescu, *USA*
Ana Maria da Silva, *Canada*
Andreas Dittrich, *Germany*
Rui M.L. Ferreira, *Portugal*
Massimo Greco, *Italy*
Willi H. Hager, *Switzerland*
Hendrik Havinga, *The Netherlands*
Hans-B. Horlacher, *Germany*
David Hurther, *France*
Martin Jäggi, *Switzerland*
Juha Järvelä, *Finland*
Sameh Kantoush, *Egypt*
Katinka Koll, *Germany*
Bommanna G. Krishnappan, *Canada*
Stuart Lane, *Switzerland*
João G.A.B. Leal, *Norway*
Angelo Leopardi, *Italy*

Danxun Li, *China*
Juan-Pedro Martín-Vide, *Spain*
Bijoy S. Mazumder, *India*
Bruce W. Melville, *New Zealand*
Emmanuel Mignot, *France*
Rafael Murillo, *Costa Rica*
Heidi Nepf, *USA*
A. Salehi Neyshabouri, *Iran*
Vladimir Nikora, *UK*
Helena I.S. Nogueira, *Portugal*
Nils Reidar Olsen, *Norway*
André Paquier, *France*
Piotr Parasiewicz, *USA*
Michel Pirotton, *Belgium*
Dubravka Pokrajac, *UK*
Sebastien Proust, *France*
Wolfgang Rodi, *Germany*
Jose Rodriguez, *Australia*
Pawel M. Rowinski, *Poland*
André Roy, *Canada*
Koji Shiono, *UK*
Graeme M. Smart, *New Zealand*
Sandra Soares-Frazão, *Belgium*
Thorsten Stoesser, *UK*
Mutlu Sumer, *Denmark*
Simon Tait, *UK*
Aldo Tamburrino, *Chile*
Wim S.J. Uijttewaal, *The Netherlands*
Zhaoyin Wang, *China*
Volker Weitbrecht, *Switzerland*
Silke Wieprecht, *Germany*
Farhad Yazdandoost, *Iran*
Yves Zech, *Belgium*

MEMBERS OF THE IAHR COMMITTEE ON FLUVIAL HYDRAULICS

Andreas Dittrich, *Technical University of Braunschweig, Germany, Chair*
Jose Rodríguez, *The University of Newcastle, Australia, Vice Chair*
Mustafa Altınakar, *The University of Mississippi, USA, Past Chair*
André Paquier, *Irstea, France, Member*
Sandra Soares-Frazão, *Université Catholique de Louvain, Belgium, Member*
Subhasish Dey, *Indian Institute of Technology, India, Member*
Angelo Leopardi, *Università Degli Studi Di Cassino E Del Lazio Meridionale, Italy, Member*
Shaohua Marko Hsu, *Feng Chia University, Taiwan, Member*
Rui M.L. Ferreira, *Instituto Superior Técnico, Portugal, Co-opted Member*
George Constantinescu, *University of Iowa, USA, Co-opted Member*
Mário J. Franca, *École Polytechnique Fédérale de Lausanne, Switzerland, Co-opted Member*

MEMBERS OF THE LOCAL ORGANIZING COMMITTEE

Anton J. Schleiss, *EPFL, Conference Chairman*
Giovanni De Cesare, *EPFL, Co-Chairman*
Mário J. Franca, *EPFL, Co-Chairman*
Michael Pfister, *EPFL, Co-Chairman*
Scarlett Monnin, *EPFL, Secretary*
Gesualdo Casciana, *EPFL, Secretary*

Reservoir Sedimentation – Schleiss et al. (Eds)
© 2014 Taylor & Francis Group, London, ISBN 978-1-138-02675-9

Sponsors

SUPPORTING INSTITUTIONS

ÉCOLE POLYTECHNIQUE
FÉDÉRALE DE LAUSANNE

SWISS NATIONAL SCIENCE FOUNDATION

Schweizerischer Wasserwirtschaftsverband
Association suisse pour l'aménagement des eaux
Associazione svizzera di economia delle acque

GOLD SPONSORS

Schweizerische Eidgenossenschaft
Confédération suisse
Confederazione Svizzera
Confederaziun svizra

Swiss Confederation

Federal Office for the Environment FOEN

SILVER SPONSORS

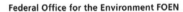
Basler & Hofmann

MET-FLOW

IM Engineering | **IUB** Engineering

**AquaVision
Engineering**

Reservoir sedimentation

Tourtemagne reservoir (VS), Switzerland.

Reservoir Sedimentation – Schleiss et al. (Eds)
© *2014 Taylor & Francis Group, London, ISBN 978-1-138-02675-9*

Sustainable water supply, climate change and reservoir sedimentation management: Technical and economic viability

G.W. Annandale

Golder Associates Inc., Denver, Colorado, USA

ABSTRACT: The paper considers the effects of climate change and reservoir sedimentation on water supply reliability and sustainability. From a global fresh water supply perspective it is argued that river water has the greatest potential for sustainable development. The known fact that reservoir storage is required to reliably supply water from rivers leads to the conclusion that provision of such space will become even more important in the future due to the uncertainties of increased hydrologic variability associated with climate change. The limited availability of dam sites emphasizes the need to indefinitely maintain reservoir storage space for use by future generations, thereby ensuring sustainable development. The current undesirable situation that is characterized by more reservoir storage space being lost to sedimentation than is added through construction of new facilities points to the importance of implementing reservoir sedimentation management techniques. Historically, its implementation has been hampered by an incorrect perception that reservoir sedimentation management is not economically viable. To transform this view it is necessary to acknowledge the dual nature of reservoir storage, i.e., it can be either renewable or exhaustible depending on design and operating decisions. Correct characterization of reservoir storage space and implementation of principles of the economics of exhaustible resources quantifies the value to reservoir sedimentation management, previously ignored. Sustainable and reliable supply of fresh water in the future demands changes in engineering and operating paradigms and changes in the ways dams and their reservoirs are economically evaluated.

1   INTRODUCTION

Global fresh water supply is in crisis, due to increasing world population, non-sustainable development and use of water resources and the imminent threat associated with climate change. This paper provides a high-level review of the potential sustainability of abundantly available fresh water sources and identifies the preferred source for future development, i.e., river water. It confirms the need for storage to reliably supply fresh water from rivers and clearly demonstrates that the need for large reservoir storage spaces will increase in the future to reliably supply fresh water from rivers under climate change scenarios. The limited availability of suitable dam sites to develop the required reservoir storage space and the threat of reservoir sedimentation, which reduces available reservoir storage space, emphasizes the importance of sustainable development. With reservoir sedimentation being the single greatest threat to losing reservoir storage space, the importance of reservoir sedimentation management is emphasized. Indefinitely preserving reservoir storage space for use by future generations is of critical importance. Viable reservoir sedimentation management techniques exist, but their implementation has been hampered in the past due to the incorrect perception that such management approaches are not economically viable. This notion originates from standard discounting procedures used in economic analysis, which merely represents current cultural values but is not in compliance with the principles of sustainable development, i.e., a desire for fairness between current and future generations—a concept known as creation of intergenerational equity. By highlighting the dual nature of reservoir storage space it is argued that correct use of the principle of the economics of exhaustible resources results in decisions favoring the use of reservoir sedimentation management technology to ensure sustainable development of water resource infrastructure.

2 SUSTAINABLE FRESH WATER SOURCES

The two principal and most abundantly available sources of fresh water are groundwater and river water. Identification of the source with the greatest potential for sustainable development requires consideration of principles proposed by Herman Daly, an economist who participated in the development of sustainable development policy guidelines for the World Bank. He proposed for sustainable development to occur that a) renewable resources should be used at a *rate* that is smaller than their *rate* of regeneration, b) exhaustible resources should be used at a *rate* that is smaller than the *rate* of development of renewable substitutes, and that c) pollution should not exceed the *rate* by which the environment can assimilate it.

Due to the hydrologic cycle both groundwater and surface water may be viewed as renewable and both have the potential for sustainable development if the *rate* of usage does not exceed the *rate* of replenishment. Globally, it has been found that groundwater is not sustainably developed nor used. Gleeson et al. (2012) found that 3.5 times more groundwater is globally used than what is replenished. Konikow (2011) sets average groundwater depletion since 1900 at about 4,500 km³, which equals the net storage space of all large reservoirs on earth. This non-sustainable use of groundwater is attributed to the large difference between global replenishment rates of groundwater (about 1,400 years) (Shiklomanov & Rodda 2003) and daily usage rates. In contrast Shiklomanov & Rodda (2003) estimates the global replenishment rate of river water at about 16 to 18 days, which is much closer to the daily rate of abstraction of water. This indicates that river water has a much greater potential for sustainable development.

3 NEED FOR STORAGE

By selecting river water as the preferred source for sustainable development of water supply systems the indispensable need for storage to reliably supply fresh water is acknowledged. Storage is required to smooth out seasonal variations in river flow, which in some places may also vary significantly from year to year. Regions with high inter-annual hydrologic variability require

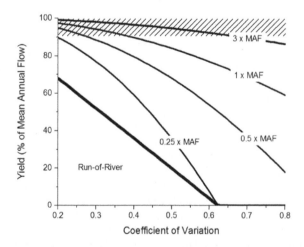

Figure 1. Relationship between yield, annual coefficient of variation of river flow (standard deviation divided by the mean) and reservoir storage volume (expressed in terms of mean annual flow—MAF) for 99% reliability of supply. Bold line separates run-of-river and multiple-year storage states (Annandale 2013).

large reservoir storage volumes to enable reliable supply of water during multiple-year droughts. Figure 1, based on the Gould-Dincer approach (McMahon et al. 2007) distinguishes between run-of-river and storage requirements, dependent on inter-annual hydrologic variability.

4   CLIMATE CHANGE AND ROBUST INFRASTRUCTURE

Global increases in hydrologic variability due to climate change (i.e., more intense and longer droughts, and greater floods) will have the greatest impact on water supply reliability. Although it is uncertain by how much hydrologic variability will increase, an impression of its impacts can be found by delineating how regions currently experiencing multiple-year droughts might expand. Figure 2(a) approximates regions of the world currently experiencing on a regular basis multiple-year droughts (two years or longer). Figure 2(b) is an estimate of how the area of such regions might expand if global hydrologic variability would increase by 25% (Annandale 2013). The figure implies that regions of the world currently using run-of-river projects or small reservoirs to reliably supply water will require much larger storage reservoirs in the future. For example, using the Gould-Dincer approach (McMahon et al. 2007), Figure 3 shows that if the annual coefficient of variation of riverflow increases from 0.4 to 0.5, then the yield will decrease from 60% of the Mean Annual Flow (MAF) to about 35% MAF for a reservoir with a volume equaling 0.25 MAF. In contrast, when using a reservoir with a volume equaling one times the MAF, the yield will decrease from 90% to about 85% of the MAF. Yield is therefore much less sensitive to increased hydrologic variability when

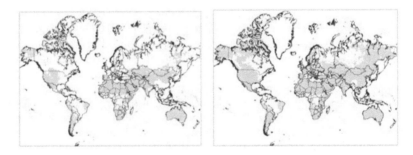

Figure 2.   (a) Estimated delineation of regions experiencing multiple-year droughts on a regular basis. (b) Estimated expansion of regions that will experience multiple-year droughts for an assumed global increase in hydrologic variability equaling 25% (Annandale 2013).

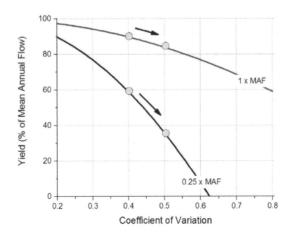

Figure 3.   Sensitivity of yield to increased hydrologic variability from reservoirs with volumes equaling 0.25 and 1.00 times the MAF (99% reliability).

using large storage spaces than smaller ones. Large reservoir volumes therefore characterizes robust infrastructure, resistant to the effects of climate change.

5 SUSTAINABLE DEVELOPMENT

Sustainable development as defined by the Brundtland Commission (United Nations 1987) requires fairness between generations, i.e., creation of intergenerational equity. To create intergenerational equity in an era of climate change will necessitate having available enough reservoir storage for use by future generations. Not having enough storage will prevent future generations from reliably supplying fresh water from rivers.

Current trends in net reservoir storage space indicate that enough storage may not be available in the future. The combined effects of a reduced global rate of dam construction since about the 1980s and storage loss to reservoir sedimentation resulted in a global net decrease in reservoir storage (Fig. 4). The only way to reverse this undesirable trend is to construct more reservoirs and to implement reservoir sedimentation management approaches that will indefinitely preserve reservoir storage space. The limited availability of suitable dam sites for construction of reservoirs emphasizes the importance of implementing reservoir sedimentation management technology in both existing and future projects.

6 RESERVOIR SEDIMENTATION MANAGEMENT

Multiple reservoir sedimentation management techniques exist and their technical viability have been proven (Morris & Fan 1997; Sumi 2003; Palmieri et al. 2003; Annandale 2011). Techniques may be categorized into groups representing catchment management, prevention of sediment deposition and removal of deposited sediment. Each of these categories contains a number of techniques, as summarized in Figure 5. Catchment management aims are reducing sediment yield, i.e., the amount of sediment that may flow into a reservoir. Techniques that may be used to accomplish this goal include reforestation (revegetation of catchments), construction of check dams, contour farming and warping. Prevention of sediment deposition in reservoirs can be accomplished through bypassing sediment around a reservoir (using tunnels, river modification, sediment exclusion and off-channel storage) and by sluicing and

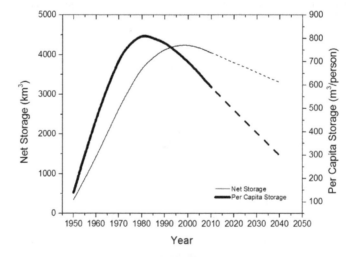

Figure 4.   Trends in net global total and per capita net reservoir storage (developed from White 2001 and assuming 1% annual loss to sedimentation).

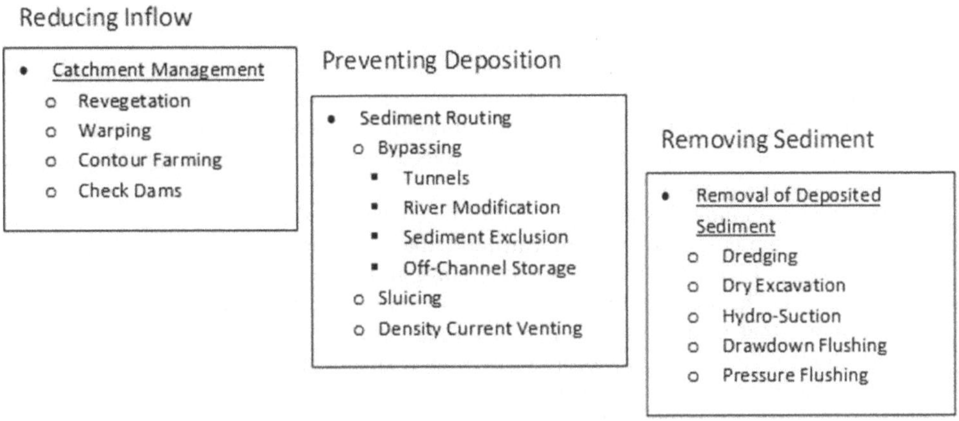

Reducing Inflow

- Catchment Management
  - Revegetation
  - Warping
  - Contour Farming
  - Check Dams

Preventing Deposition

- Sediment Routing
  - Bypassing
    - Tunnels
    - River Modification
    - Sediment Exclusion
    - Off-Channel Storage
  - Sluicing
  - Density Current Venting

Removing Sediment

- Removal of Deposited Sediment
  - Dredging
  - Dry Excavation
  - Hydro-Suction
  - Drawdown Flushing
  - Pressure Flushing

Figure 5.   Three categories of reservoir sedimentation management techniques.

density current venting. Various techniques can be used to remove sediment that has already deposited in reservoirs. These include drawdown flushing, pressure flushing, dredging, dry excavation and hydrosuction. Selecting the right technique is project specific, a topic that is beyond the scope of this paper.

7   DUAL NATURE OF STORAGE

A question requiring addressing is why reservoirs continue to lose storage space due to sedimentation in spite of the fact that multiple techniques are available to either prevent sedimentation or, at least, minimize the magnitude of storage loss. Part of the reason for this undesirable state of affairs can be found in a common design paradigm and belief that little can be done to prevent reservoir sedimentation. This attitude results in dam designs that are not amenable to reservoir sedimentation management.

Another reason is that the economic analysis techniques that are used to assess the economic value of dams and reservoirs do not promote sustainable development. Conventional discounting techniques merely reflect existing cultural values that place greater importance on the present than on the future, a position that is in conflict with the tenets of sustainable development.

To address this issue it is necessary to revert to the sustainable development principles proposed by Daly, and firstly determine whether reservoir storage space is an exhaustible or a renewable resource. As already indicated, the limited number of suitable dam sites available for construction of reservoir storage implies that reservoir storage space may be an exhaustible resource. However, closer examination shows that this is not strictly true. Reservoir storage space potentially has a dual character; it can be either exhaustible or renewable depending on design and operating decisions.

If a dam designer or operator accepts that storage loss due to sedimentation is inevitable and makes no attempt to develop sustainable designs or implement operating procedures that will either prevent or minimize storage loss to sedimentation, then the reservoir storage space is deliberately and consciously classified as an exhaustible resource. On the other hand, if a dam designer prepares designs facilitating reservoir sedimentation management and an operator implements management strategies that either prevent or minimize the effects of reservoir sedimentation, then the reservoir storage space is classified as a renewable resource.

These deliberate design and operating decisions, classifying a reservoir as either exhaustible or renewable, have significant implications for the economic analysis of dams and their reservoirs, thereby determining whether reservoir sedimentation management is economically viable.

Conventional approaches to assess the economic value of a project entail quantifying its net present value. This is accomplished by making use of discounting techniques. The net present value of a project is determined by subtracting discounted future costs from discounted future benefits. If the net present value thus determined is positive, a project may be viewed as economically viable.

By critically examining this approach it can be concluded that discounting in economic analysis merely represents current cultural attitudes that value the present more than the future. This can be seen in Figure 6, which shows the change in the present value of $1,000 over time, discounted at 6%. The fact that the present value of this amount of money rapidly decreases over time shows that greater value is placed on receiving benefits earlier rather than later. Therefore, the economic value of a project can be increased by scheduling benefits to occur as soon as possible in a project timeline and expenses as far into the future as possible.

The cost to future generations of losing a natural resource (i.e., the loss of reservoir storage space to sedimentation) is never accounted for in the economic analysis of dam and reservoir projects. The standard reasoning is that the present value of such a loss, which may occur 70 or 100 years from now, is insignificant. Such reasoning is not in line with the concept of sustainable development, i.e., it does not create intergenerational equity (United Nations 1987). Forcing future generations to bear the cost of discarded natural resources for the benefit of current generations is not in line with the desire to establish fairness between generations. The fact that reservoir storage space has a dual character and the fact that its characterization as either a renewable or exhaustible resource depends on deliberate decisions by dam designers and operators invalidates the reasoning common to standard economic analysis.

A deliberate decision by dam designers and operators to accept storage loss to reservoir sedimentation as inevitable, thereby consciously classifying the reservoir storage space as an exhaustible resource, demands implementation of the Hotelling Principle (Hotelling 1931). This fundamental principle of the economics of exhaustible resources (Solow 1974) indicates that the price of an exhaustible resource increases with the discount rate, which has significant implications to the economic analysis of dam and reservoir projects. If the price of an exhaustible resource increases with the discount rate, then the present value of that resource remains constant (Gopalakrishnan 2000). Figure 6 compares the present value of an exhaustible resource using the Hotelling Principle to the present value determined using conventional discounting. The present value of $1,000 using the Hotelling Principle remains $1,000 regardless of time.

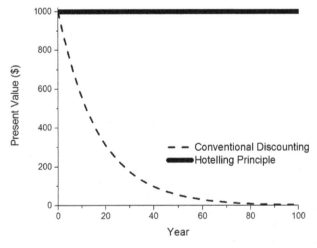

Figure 6.   The present value of $1,000 using conventional discounting techniques (for a discount rate of 6%) and using the Hotelling Principle.

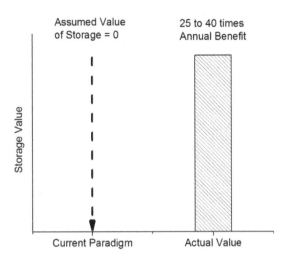

Figure 7. The value of reservoir storage space in economic analysis, based on the current paradigm and an estimate of its actual value.

Deliberate decisions by designers and operators to classify reservoir storage space as an exhaustible resource, while in fact it can be developed and managed as a renewable resource, demands recognizing in economic analysis of the cost of losing the resource to sedimentation. A simplified approach to determining its value may be to multiply the present value of indefinite use of a reservoir (assuming no storage loss to sedimentation) with an economic multiplier of dams and reservoirs. By using the range of economic multipliers found by Bhatia et al. (2008) for dams and reservoirs, i.e., between 1.40 and 2.40, it can be shown that the value of reservoir storage space may range between 25 and 40 times the annual net benefit (Annandale 2013) (Fig. 7). An important implication when recognizing the value of reservoir storage space in economic analysis and concurrently invoking the use of the Hotelling Principle is that the cost of reservoir sedimentation management is offset by maintaining the value of the reservoir. This is not the case in conventional economic analysis, where the cost of reservoir sedimentation management is deemed as resulting in no value, due to the fact that the loss of reservoir storage space is deemed of no consequence.

9   CONCLUSION

Review of the most abundantly available fresh water sources on earth, i.e., groundwater and river water, concludes that groundwater is non-sustainably used and that river water has the greatest potential for sustainable development. The use of rivers as the preferred fresh water supply source requires storage to reliably supply water. This need for storage will increase in the future due to the expected increase in hydrologic variability resulting from the effects of climate change. Investigation of the relationship between yield, reservoir storage volume and hydrologic variability reveals that robust infrastructure is defined as hydrologically large reservoir volumes. The yield from reservoirs with large volumes is less sensitive to the effects of climate change than that from reservoirs with small volumes. Therefore, to maintain the reliability of water supply in spite of the anticipated effects of climate change it is necessary to have available as much reservoir storage as possible.

Historic trends in global net reservoir storage reveal a negative drift, i.e., currently more storage space is lost to reservoir sedimentation than what is added through the construction of new facilities. This negative trend does not bode well for the future. The reduced rate of dam construction since about the 1980s and continued storage loss to reservoir sedimentation will result in a reduction in the reliability of water supply. The remedy is two-fold. Firstly, it is necessary to construct more dams to create reservoir storage space. Secondly, it is necessary to implement management strategies to either eliminate or reduce the rate of storage loss to sedimentation. Implementation of such techniques is required at both existing and future facilities. Future facilities should be designed to facilitate reservoir sedimentation management, while existing facilities should be retrofitted to accomplish the same.

To achieve these goals it is necessary to change current engineering design and economic analysis paradigms. Accepting as unavoidable storage loss to reservoir sedimentation results in a conscious decision by the designer and operator to classify reservoir storage space as an exhaustible resource. This is unnecessary. Reservoir storage space has a dual character. By making decisions to manage reservoir sedimentation and thereby either eliminating storage loss or drastically reducing the rate of storage loss, the same reservoir may be classified as a renewable resource. Therefore, if a designer or operator deliberately decides to accept storage loss to reservoir sedimentation as inevitable it is necessary to modify the economic analysis approach. This is done by recognizing the value of reservoir storage space and accounting for its loss to reservoir sedimentation by making use of the Hotelling Principle, i.e., the fundamental principle of the economics of exhaustible resources. Implementation of this principle allows quantification of the value of reservoir sedimentation management. Managing reservoir storage space as a renewable resource is of critical importance to create intergenerational equity, thereby accomplishing the goals of sustainable development and addressing the uncertainties associated with climate change. Designing and operating reservoirs following a life-cycle management approach aimed at indefinitely preserving reservoir storage space is significant for the continued survival of man.

REFERENCES

Annandale, G. 2013. *Quenching the Thirst—Sustainable Water Supply and Climate Change*, CreateSpace, Charleston, SC.

Annandale, G.W. 2011. Going Full Circle, *International Water Power and Dam Construction*, April, pp. 30–34.

Annandale, G.W. 1986. *Reservoir Sedimentation*, Elsevier Science Publishers, Amsterdam.

Basson, G.R. & A. Rooseboom. 1999. *Dealing with Reservoir Sedimentation*, Water Research Commission, Pretoria, South Africa.

Bates, B.C. et al. Eds. 2008.—Climate Change and Water. *Technical Paper of the Intergovernmental Panel on Climate Change*, IPCC Secretariat, Geneva, 210 pp.

Bhatia, R. et al. 2008. *Indirect Economic Impacts of Dams: Case Studies from India, Egypt and Brazil*, The World Bank, Washington D.C.

Gleeson, T. et al. 2012. Water balance of global aquifers revealed by groundwater footprint, *Nature*, Vol. 488, August 9, pp. 197–200.

Gopalakrishnan, C. (Ed). 2000. *Classic Papers in Natural Resource Economics*, MacMillan Press Ltd, London.

Hotelling, H.M. 1931. The Economics of Exhaustible Resources, *Journal of Political Economy*, Vol. 39, pp. 137–75.

Konikow, L.F. 2011. Contribution of global groundwater depletion since 1900 to sea-level rise, *Geophysical Research Letter*, Vol. 38, L1749, doi:10.1029/2011 GL048604.

McMahon, T.A. et al. 2007. Review of Gould-Dincer Reservoir Storage-Yield-Reliability Estimates, *Advances in Water Resources*, Vol. 30, pp. 1873–1882.

Morris, G.M. & 1997. *Reservoir Sedimentation Handbook*, McGraw-Hill, New York.

Palmieri, A. et al. 2003. *Reservoir Conservation: The RESCON Approach*, World Bank, Washington D.C.

Shiklomanov, I.A. & Rodda, J.C. 2003. *World Water Resources at the Beginning of the 21st Century*, International Hydrology Series, Cambridge University Press.

Solow, R.M. 1974. The Economics of Resources or the Resources of Economics, *American Economic Review*, Vol. 64, pp. 1–15.

Sumi, T. 2003. *Approaches to Reservoir Sedimentation Management in Japan, Reservoir Sedimentation Management Symposium*, Third World Water Forum, Kyoto, Japan.

United Nations. 1987. *Our Common Future, Report of the World Commission on Environment and Development Chapter 1: A Threatened Future, Paragraph 49*, Report Number A/42/427, New York.

White, R., 2001. *Evacuation of sediments from reservoirs*, Thomas Telford Publishing, London, United Kingdom, 280 pp.

Reservoir Sedimentation – Schleiss et al. (Eds)
© 2014 Taylor & Francis Group, London, ISBN 978-1-138-02675-9

Three-dimensional numerical modeling of flow field in rectangular shallow reservoirs

T. Esmaeili & T. Sumi
Department of Urban Management, Graduate School of Engineering, Kyoto University, Kyoto, Japan

S.A. Kantoush
Department of Civil Engineering, German University in Cairo (GUC), Cairo, Egypt

A.J. Schleiss
Ecole Polytechnique de Lausanne (EPFL), Laboratory of Hydraulic Constructions (LCH), Lausanne, Switzerland

S. Haun
Norwegian University of Science and Technology (NTNU), Trondheim, Norway

ABSTRACT: Flow field in shallow waters, which is characterized by its complex mixing process and inherent dynamic nature, is interesting mainly due to its practical importance (e. g. in free flushing operation and sedimentation in large reservoirs). 3D numerical models make it possible to track two-dimensional large turbulence coherent structures, which are the dominant phenomenon in shallow reservoirs flow field. In the present study a fully three-dimensional numerical model SSIIM that employs the Finite Volume Approach (FVM) was utilized to reproduce the 3D flow field. Various shallow reservoir geometries with fixed and deformed equilibrium bed were considered. The measurements by Large-Scale Particle Image Velocimetry techniques (LSPIV) and Ultrasonic Doppler Velocity Profiler (UVP) over the flow depth were used for model validation. Outcomes revealed reasonable agreement between the simulated and measured flow velocity field even when an asymmetric flow pattern exists in the reservoir.

1 INTRODUCTION

Shallow waters are defined as a flow field in which the vertical dimension of fluid domain is significantly smaller than its horizontal dimensions (Yuce & Chen 2003). Wide rivers, lakes, coastal lagoons, estuaries as well as large reservoirs are the examples of shallow waters in the real life. Flow pattern in wide and shallow reservoirs with sudden expansion of inlet section may become unstable, which produces large-scale transverse motions and recirculation zones due to the transverse disturbance because of the high sensitivity of flow pattern to the initial and boundary condition (Dewals et al. 2008). This type of flows is prominent in the nature and also emerges in various engineering applications including sudden expansions (Shapira et al. 1990), compound channels (Ghidaoui & Kolyshkin 1999 and Chu et al. 1991), storage chambers (Adamson et al. 2003), settling tanks (Frey et al. 1993) as well as shallow reservoirs sedimentation (Kantoush et al. 2008a & 2010).

When large-scale transverse motions and turbulent coherent structures emerge in shallow reservoirs, the sediment transportation pattern would be seriously affected by the flow velocity field. Subsequently, measurement of 2D surface velocity and vertical velocity components with high spatial resolution is essential to predict the favorable sedimentation zone. Such kind of knowledge leads to more efficient sediment management strategies in reservoirs. Also, assessment of the flow field is necessary to characterize the domain of main jet flow, reverse flow and eddies within a shallow reservoir.

Kantoush (2007) presented a comprehensive review of experimental tests in shallow reservoirs with transverse flow motions in symmetric channel expansions. The observations revealed that asymmetric flow pattern can be developed under a certain geometric and hydraulic condition even if the symmetric geometry and hydraulic condition is employed. Same outcome has been obtained by Adamson et al. (2003) and Stovin & Saul (1996) regarding the storage tank sedimentation and storage chambers, respectively. However, most of the studies in the literature considered the sudden plane expansions of an infinite length and studies focusing on various geometric and hydraulic parameters are very limited.

Mizushima & Shiotani (2001) used numerical model for studying the flow instabilities in symmetric channels with sudden expanded section. Dewals et al. (2008) and Dufresne et al. (2011) used 2D numerical models to investigate the flow pattern distribution in shallow reservoirs. Nonetheless, the one and two dimensional models are not able to directly simulate the secondary current influences in complex 3D flows, especially on deformed beds, since the complexity of flow pattern is further magnified.

In the present study, four different reservoir geometries with different length-to-width (Aspect ratio that is called AR hereafter) and reservoir width to inlet channel width (Expansion ratio that is called ER hereafter) were considered for numerical simulation. The fixed bed and also equilibrium bed after free-flow flushing were used as the initial bed condition as well. Numerical results were compared with the experimental measurements of surface velocity for all cases as well as 3D velocity component for one case and a satisfactory agreement was found between the predicted and observed flow pattern. Furthermore, the numerical model could reproduce the steady asymmetric flow pattern when a perturbation quantifier was introduced.

2 MATERIAL AND METHODS

2.1 *Experimental setup and conditions*

The experimental tests were carried out at the Laboratory of Hydraulic Constructions of Swiss Federal Institute of Technology (EPFL) in a rectangular reservoir with the maximum inner length (L) of 6 m and width (B) of 4 m. Also, the inlet and outlet rectangular channel width (b) and length (l) were 0.25 m and 1 m respectively. Both channels were installed at the center of upstream and downstream side wall of the reservoir. The different shallow reservoir geometry achieved experimentally by moving the PVC plate walls. The reservoir depth is 0.3 m and the both side walls and bottom is hydraulically smooth and flat. The water level in the reservoir was controlled by a 0.25 m width and 0.3 m height flap gate set up at the end of outlet channel. A moveable frame with 4 m length was mounted on the side walls of the reservoir for installing the measurement devices. Table 1 shows geometrical attributes of four geometries employed in the present study. As for the geometrical parameters, ER = B/b shows the influence of change in the reservoir width while AR = L/B is appropriate for describing the effect of variations in the reservoir length (Dewals et al. 2008).

Large Scale Particle Image Velocimetry technique (LSPIV) was used for measuring the surface velocity field. Ultra Sonic Velocity Profiler device (UVP) was employed for providing the 3D flow velocity measurements as well. Also, within the frame work of experimental study, sedimentation and sediment flushing from the shallow reservoirs were investigated (Kantoush et al. 2008b). Non-uniform crushed walnut shells were used. The median size of this non-cohesive light-weight and homogenous grain material was 50 μm with σ_g of 2.4 and a density of 1500 kg/m³. The flow discharge rate (Q) and water depth (h) were constant for all experiments as 0.007 m³/s and 0.2 m respectively except for the cases with deformed equilibrium bed. Thus, in all tested configurations with fixed bed, the measured Froude and Reynolds number at the inlet channel were kept constant as Fr = 0.1 and Re = 1.75 × 10⁴. In case of measurements of flow field on deformed bed after flushing the water level and discharge were 0.1 m and 0.007 m³/s respectively.

Figure 1 illustrates the observed flow streamlines for case T8 and T13. As can be observed, the issuing flow jet deviated to the right hand side which forms asymmetric flow pattern in

Table 1. Experimental configuration and corresponding geometrical parameters.

Case	L(m)	B(m)	Pr(m)*	ER(–)	AR(–)	SF(–)**
T8	6.0	2.0	15.5	8.0	3.0	0.375
T9	6.0	1.0	13.5	4.0	6.0	0.122
T11	5.0	4.0	17.5	16.0	1.25	0.99
T13	3.0	4.0	13.5	16.0	0.75	0.97

* Pr is the wetted perimeter.
** A is the total reservoir area and SF is the shape factor which defined as (A/Pr2) × ER.

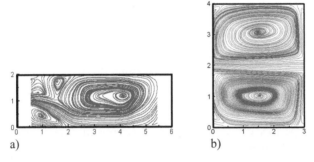

a)   b)

Figure 1.   Observed streamlines of the surface flow field for a) case T8 and b) case T13.

case T8. The main eddy rotating anticlockwise in the center part of the reservoir and two others rotate clockwise in the upstream corners. Also, symmetric flow pattern with one main jet trajectory in the centerline and two side eddies has been developed for case T13. Kantoush (2007) concluded that the deviation to the right hand side is due to the random disturbance of the initial boundary condition and a mirror situation would be easily established by disturbing the initial condition slightly. The flow deviation to one side of the reservoir corresponds to the increase of flow velocity in one side of the jet and consequent reduction of the pressure. This process will lead to flow deviation to one side of the reservoir and called Coanda effect (Chiang et al. 2000).

2.2   Numerical model & features

Fully 3D numerical model SSIIM was employed in this study. The numerical model solves the mass conservation and Reynolds-averaged Navier-Stokes equation in three dimensions (Equation 1 and 2) to compute the water motion for turbulent flow as follows.

$$\frac{\partial U_i}{\partial x_i} = 0 \tag{1}$$

$$\frac{\partial U_i}{\partial t} + U_j \frac{\partial U_i}{\partial x_j} = \frac{1}{\rho} \frac{\partial}{\partial x_j} \left(-P\delta_{ij} + \rho \upsilon_T \left(\frac{\partial U_i}{\partial x_j} + \frac{\partial U_j}{\partial x_i} \right) \right) \tag{2}$$

in which $i = 1, 2, 3$ is the representative of three directions; where U_i is the averaged velocity, x is the spatial geometrical scale, ρ is the water density, P is the Reynolds-averaged pressure, δ_{ij} is the Kronecker delta and v_T is the turbulent eddy-viscosity. For transforming the partial equations into algebraic equations, the finite volume method is applied as discretization method, together with the second order upwind scheme.

13

The change in water-levels was based on calculated pressure field. The pressure was extrapolated to the water surface and the pressure difference between a surface node and the downstream node was used to estimate the water elevation difference (Olsen 2013). The turbulence is modeled by the standard k-ε model, using constant empirical values (Launder & Spalding 1972). The unknown pressure field is calculated employing Semi Implicit Method for Pressure-Linked Equations, (SIMPLE) method (Patankar 1980). The grid is adaptive and moves with change in the bed and water levels.

The Dirichlet boundary condition for the water inflow (logarithmic velocity distribution) was used while for the water and sediment outflow zero-gradient boundary condition was specified. For the boundary condition at the walls, where there is no water flux, the empirical wall laws introduced by Equation 3 were used:

$$\frac{U}{u^*} = \frac{1}{\kappa} \ln\left(\frac{30y}{k_s}\right)$$

(3)

where the shear velocity is denoted u^*, κ is the Karman constant equal to 0.4, y is the distance to the wall and k_s is the equivalent roughness.

The sediment transport computation for simulating the morphological change is divided into suspended and bed load transport. Suspended load is calculated by solving the transient convection-diffusion equation formula and the bed load is simulated by Van Rijn formula (Van Rijn 1984a). In order to compute the suspended sediment concentration in the cells close to the bed, a specified concentration was used as boundary condition (Van Rijn 1984b). Also, the turbulent diffusivity is calculated by Equation 4.

$$\Gamma_T = \frac{v_T}{Sc}$$

(4)

where Sc is the Schmidt number representing the ratio of eddy viscosity coefficient v_T to diffusion coefficient and is set to 1.0 as the default value in SSIIM model (Olsen 2013).

Based on the experimental measurements computational mesh for all reservoir geometries were made. The mesh cell size for case T8, T9, T11 and T13 in X and Y direction was 5 cm × 1.5 cm, 5 cm × 2.5 cm, 5 cm × 2 cm and 2.5 cm × 1 cm respectively. Considering the 11 cells for vertical grid distribution, the total number of cells over the main reservoir geometry were 174460, 52800, 220000 and 528000 respectively.

3 RESULTS AND DISCUSSIONS

3.1 Model calibration

As for the real cases in prototype scale, flow field modeling will provide us useful information about the areas with the potential erosion and deposition (sedimentation) during the anticipated floods. The more accurate the prediction of sedimentation zone, the higher is the efficiency of sediment management strategies.

In the numerical simulation, the time step was calibrated as 2 seconds for run T8 and T9 whereas it was 0.5 second for T11 and T13. In case of flow field modeling on the equilibrium deformed bed, the bed roughness was fixed as 0.00015 m which equals to 3 times the median sediment size.

3.2 Flow field modeling on fixed beds

Simulations have been conducted by employing the geometry and inflow/outflow condition of the experimental model. The k-ε turbulence model was utilized and numerical runs were performed until a steady-state flow condition is obtained. Simulations revealed that the model was not able to reproduce an asymmetric flow pattern when the geometry configuration and hydraulic boundary condition are perfectly symmetric. This is because of the implemented mathematical algorithms that were not aimed to reproduce such kind of artificial numerical outputs.

Due to the fact that symmetric flow pattern was not observed in the experiments for special geometries, Dewals et al. (2008) introduced a slight disturbance in the initial boundary condition for 2D numerical simulations. They employed non-uniform cross-sectional discharge in the inlet boundary to test the stability of the numerical model outputs. In the present study, the same concept of slight disturbance in the inlet boundary condition was implemented for all runs and non-uniform cross-sectional velocity distribution was utilized in the inflow boundary condition. Therefore, the initial velocity magnitude differ 2.5% in one side of the inlet channel comparing to the other side. Such kind of disturbance is inevitable in the experimental set up. Nonetheless, very small perturbation of the inflow condition will impose significant effect on the flow field of case T8 and T9 in the numerical model. The reason can be attributed to the unstable nature of symmetric flow in such geometries which pronounces the high sensitivity of the flow pattern to the inflow boundary condition.

Figure 2 shows the simulated and measured surface velocity magnitudes (V) in m/s and flow distribution pattern. As can be clearly seen from Figure 2, asymmetric flow pattern has developed in case T8 and T9 whereas symmetric flow pattern is observed for case T11 and T13.

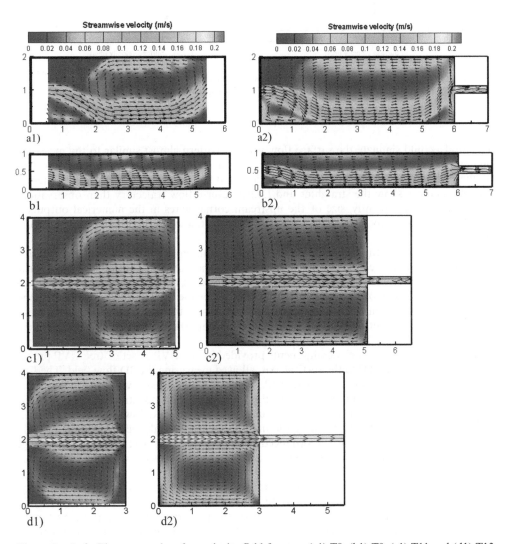

Figure 2.   Left: The measured surface velocity field for runs (a1) T8, (b1) T9, (c1) T11 and (d1) T13 respectively and right: corresponding simulated velocity field.

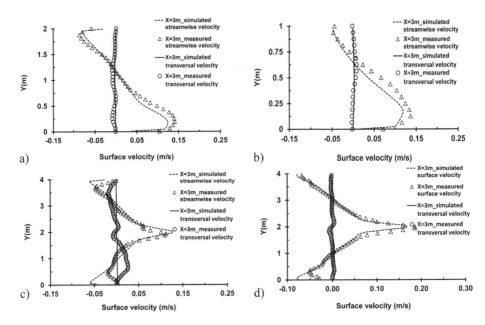

Figure 3. The measured streamwise and transversal surface velocity at the middle cross section of the reservoirs versus simulated surface velocity for (a) T8, (b) T9, (c) T11 and (d) T13.

The model could simulate the surface flow velocity pattern almost similar to the measured one by reproducing the dominant aspects such as the main flow jet trajectory and location of the reverse flow as well as the main vortices and corner gyres. Nevertheless, numerical model outcomes show the straighter and longer reverse flow trajectory than observations for all cases. Consequently, size of the upstream corner gyres in the numerical outputs is smaller than in the experimental measurements. This situation is predominant for case T11 (Figure 2c1 and 2c2).

Figure 3 illustrates the simulated streamwise and transversal surface velocity distribution versus the measured one in the middle cross-section of the reservoirs. Although the numerical model results are globally consistent with the measurements, there is a discrepancy for transversal surface velocity of case T11. The reason could be the concentrated flow pattern and lower diffusion of the main jet as well as the reverse flow in the numerical model outcomes.

The vertical distribution of flow velocity field is important for analyzing the sediment transport in reservoirs. Thus, the numerically simulated 3D flow velocity field was compared with the measured 3D velocity components provided by the UVP. A set of tree UVP probes which were inclined at 20° to the vertical axis, allowed measuring the 3D flow field. The first valid UVP measurements were located at 12.5 cm away from the side walls and 2.5 cm from the free water surface (Kantoush, 2008b). Figure 4a and 4b demonstrates the longitudinal velocity distribution over the flow depth, at upstream, middle, and downstream area of case T9. It can be seen that the higher longitudinal velocity has been deflected towards the right bank side and the reverse flow is reproduced beside the left bank side. Such kind of change in the flow direction across the reservoir is also qualitatively consistent with the experimental observations.

Velocity distribution in longitudinal, lateral and vertical directions (U, V, W respectively) at x = 5.5 m and y = 0.375 m from the right bank of the case T9 has been demonstrated in Figure 4c in order to quantitative assessment of the numerical model outputs. As for the longitudinal velocity, the numerical model overestimates the magnitudes compared with the experimental measurements while the lateral velocity was underestimated. This condition was also found for other locations over the reservoir depth. Furthermore, the vertical velocity

16

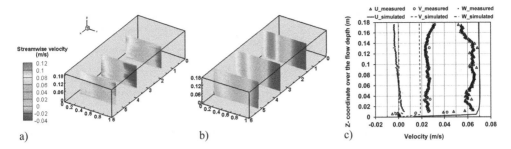

a)   b)   c)

Figure 4.   a) The UVP measurements of the longitudinal velocity over the flow depth at x = 1.5 m, 3.5 m and 5.2 m; b) the corresponding simulated velocity; c) measured and simulated velocity profiles in three dimensions: longitudinal (U), lateral (V), Vertical (W) at x = 5.5 m and y = 0.375 m position.

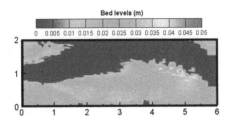

Figure 5.   The final equilibrium deformed bed after flushing for case T8.

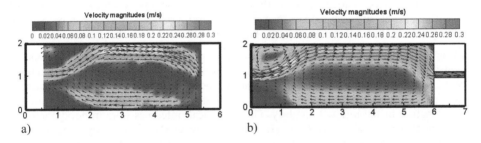

a)   b)

Figure 6.   Surface velocity field with vectors after flushing for case T8; a) measured and b) simulated.

over the flow depth remains fixed in the numerical outputs whereas there was a slight change in the measurements. However, magnitudes of the vertical velocity were very small compare to the longitudinal and lateral velocities.

3.3   *Flow field modeling on equilibrium deformed bed*

The final bed morphology, which was obtained after the sediment flushing, was introduced to the model as the boundary condition and then three-dimensional flow field was calculated. Such kind of flow field simulation is more complex than flow field on fixed beds due to the shallower flow condition with higher velocity components on the existing friction of varying bed forms.

    Figure 5 reveals the final deformed bed after sediment flushing in case T8. The figure shows that the flushing channel has shifted to the left side of the reservoir. Figure 6a and 6b also show the measured surface velocity after flushing by LSPIV technique and simulated one respectively. As can be clearly observed, the model reproduced the surface velocity pattern very similar to the observed one. Likewise the surface velocity pattern on fixed beds, the

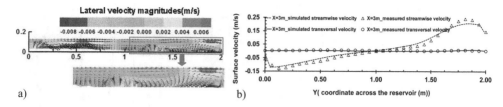

Figure 7. a) Lateral velocity distribution over the depth at x = 3 m for T8 and b) comparison of surface velocity components from the LSPIV measurements and simulations.

reverse flow trajectory is longer, straighter and also the reattachment length is smaller than the observations. However, slight disturbance in the boundary condition was not considered in the model for the current case.

Figure 7a clearly illustrates the lateral flow velocity (V) contours over the depth along with the secondary flow velocity vectors for case T8 at the middle cross-section. The streamwise and transversal simulated velocities were also plotted against the measured one in Figure 7b and good agreement was found.

4 CONCLUSIONS

In the present study, numerical model SSIIM was utilized for reproducing the 3D velocity field in rectangular shallow reservoirs with different geometries on fixed and deformed bed as well. Measured surface velocity utilizing the LSPIV technique was used for model validation. In addition, 3D velocity components over the flow depth provided by UVP compared to the numerical results. The following outcomes have been obtained from the present work:

1. Many hydrodynamic aspects such as jet trajectory, recirculation zones, eddies and the flow distribution pattern, in different shallow reservoir geometries were represented by the numerical model. The numerical model also reproduced both symmetric and asymmetric flow pattern in symmetric geometry setup, similar to the observations, by introducing a perturbation in the inflow boundary. The numerical outcomes were in good agreement with the experimental observations. However, there was some discrepancy for reproducing the upstream vortex dimension due to the longer concentrated flow and reverse flow jet pattern with lower flow diffusion in the numerical outcomes.
2. The numerical model could successfully show the effect of geometry on the flow pattern. More specifically, the numerical model showed that the flow pattern in geometries with higher SF (T11 and T13) is insensitive to the small disturbance in the inflow condition and the numerical model converged to the steady symmetric flow pattern. On the contrary, calculated flow pattern in geometries with lower SF (T8 and T9) converged to the steady asymmetric flow pattern due to the slight disturbance in the inflow boundary condition.
3. From the practical point of view, 3D numerical models can be used for reproducing the symmetric and asymmetric flow pattern under the various hydraulic and geometric conditions. Consequently, it would be possible to model the flow and sedimentation pattern as well as the effects of diverse measures on the flow field for conducting the sedimentation in preferential zones. As to the reservoirs, the outcome can be the effective prediction and management of flow field and sediment flushing process during drawdown operation (i.e. Dashidaira reservoir in the Kurobe River, Japan). Therefore, application of 3D numerical modeling beside the experimental studies result in more efficient sediment management strategies in reservoirs. However, the coupled simulation of flow and sediment field interacting with the movable bed during the drawdown flushing would be a challenging work.

REFERENCES

Adamson, A. Stovin, V. & Bergdahl, L. 2003. Bed shear stress boundary condition for storage tank sedimentation. *Journal of Environmental Engineering, ASCE* 129(7): 651–658.

Chiang, T.P. Sheu, T.W.H. & Wang, S.K. 2000. Side wall effects on the structure of laminar flow over plane symmetric sudden expansion. *Computers & Fluids* 29(5): 467–492.

Chu, V.H. Wu, J-H. & Khayat, R.E. 1991. Stability of transverse shear flows in shallow open channels. *Journal of Hydraulic Engineering, ASCE* 117(10): 1370–1388.

Dewals, B.J. Kantoush, S.A. Erpicum, S. Pirotton M. & Schleiss, A.J. 2008. Experimental and numerical analysis of flow instabilities in rectangular shallow basins. *Environmental fluid mechanics* 8(1): 31–54.

Dufresne, M. Dewals, B. Erpicum, S. Archambeau, P & Pirotton M. 2011. Numerical investigation of flow patterns in rectangular shallow reservoirs. *Engineering Application of Computational Fluid Mechanics* 5(2): 247–258.

Frey, P. Champagne, J.Y. Morel, R. & Gay, B. 1993. Hydrodynamics fields and solid particle transport in a settling tank. *Journal of Hydraulic Research,* 31(6): 736–776.

Ghidaoui, M.S. & Kolyshkin, A.A. 1999. Linear stability of lateral motions in compound open channels. *Journal of Hydraulic Engineering, ASCE* 125(8): 871–880.

Kantoush, S.A. 2007. Symmetric or asymmetric flow patterns in shallow rectangular basins with sediment transport. JFK paper competition, In International Association of Hydraulic Engineering and Research (ed.), *32rd IAHR Congress: Harmonizing the Demands of Art and Nature in Hydraulics; Proc. intern. conf., Italy, 1–6 July 2007*. Venice.

Kantoush, S.A. Bollaert, E. & Scleiss, A.J. 2008a. Experimental and numerical modelling of sedimentation in a rectangular shallow basin. *International Journal of Sediment Research,* 23(3): 212–232.

Kantoush, S.A. De Cesare, G. Boillat, J.L. & Schleiss, A.J. 2008b. Flow field investigation in a rectangular shallow reservoir using UVP, LSPIV and numerical modelling. *Flow Measurement & Instrumentation,* 19: 139–144.

Kantoush, S.A. Sumi, T. & Schleiss, A.J. 2010. Geometry effect on flow and sediment deposition pattern in shallow basins. *Annual Journal of Hydraulic Engineering, JSCE* 54: 212–232.

Launder, B.E. & Spalding, D.B. 1972. Lectures in mathematical models of turbulence. London: Academic press.

Mizushima, J. & Shiotani, Y. 2001. Transitions & instabilities of flow in a symmetric channel with a suddenly expanded and contracted part. *Journal of Fluid Mechanics* 434: 355–369.

Olsen, N.R.B. 2013. *A three dimensional numerical model for simulation of sediment movement in water intakes with multiblock option.* Department of Hydraulic and Environmental Engineering, The Norwegian University of Science and Technology, Trondheim.

Patankar, S.V. 1980. Numerical heat transfer and fluid flow. New York: McGraw-Hill.

Shapira, M. Degani, D. & Weihs, D. 1990. Stability and existence of multiple solutions for viscous flow in suddenly enlarged channels. *Computers & Fluids* 18(3): 239–258.

Stovin, V.R. & Saul, A.J. 1996. Efficiency prediction for storage chambers using computational fluid mechanics. *Water Sciences & Technology* 33(9): 163–170.

Van Rijn, L.C. 1984a. Sediment transport. Part I: Bed load Transport. *Journal of Hydraulic Engineering, ASCE* 110(10): 1733–1754.

Van Rijn, L.C. 1984b. Sediment transport. Part II: Suspended load Transport. *Journal of Hydraulic Engineering, ASCE* 110(11): 1431–1456.

Yuce, M.I. & Chen, D. 2003. An experimental investigation of pollutant mixing and trapping in shallow costal re-circulating flows. In Uijttewaal, W.S.J. & Gerhard, H. J (ed.), *Shallow flows; Proc. intern. symp., Netherlands, 16–18 June 2003*. Delft: Balkema.

Reservoir Sedimentation – Schleiss et al. (Eds)
© *2014 Taylor & Francis Group, London, ISBN 978-1-138-02675-9*

Meandering jets in shallow rectangular reservoir

Y. Peltier, S. Erpicum, P. Archambeau, M. Pirotton & B. Dewals
ArGEnCo Department, Research Group Hydraulics in Environmental and Civil Engineering (HECE), University of Liège, Liège, Belgium

ABSTRACT: In this article, meandering flows in a shallow rectangular reservoir are experimentally and numerically investigated. Two experiments were performed in a smooth shallow horizontal flume and the surface velocity fields were measured by Large-Scale PIV (LSPIV). The flow conditions were chosen in such a way that the friction regime of both flows was different. These flows were then modelled using the academic code WOLF2D, which solves the 2D shallow water equations and uses a depth-averaged k-ε model for modelling turbulence. The main characteristics of the measured and simulated flows were finally extracted from a Proper Orthogonal Decomposition (POD) of the surface velocity fields and depth-averaged velocity fields respectively and were compared. When the mean fluctuating kinetic energy of the considered POD mode is greater than 1×10^{-5} m²/s², the numerical modelling and the experiments are in good agreement whatever the friction regime of the flow.

1 INTRODUCTION

Shallow rectangular reservoirs are common structures in the field and they are used for water storage or sediment trapping. Their optimal exploitation strongly depends on the accurate prediction of the flow patterns that develop inside. Previous studies emphasized that, depending on the shape factor, $\mathrm{SF} = L/\Delta B^{0.6} b^{0.4}$ (L the reservoir length, b the width of the inlet channel and ΔB the width of the sudden expansion), and on the Froude number at the inlet, $\mathrm{F} = U_{in}/\sqrt{gH}$ (U_{in} the velocity at the inlet, H the mean water depth in the reservoir), different types of flow occur (Peltier *et al.*, 2014a):

- Symmetric with a straight central jet (SF ≤ 6.2 and F < 0.21),
- Asymmetric with the central jet impinging one or several times the lateral wall of the reservoir (SF ≥ 8.1).
- Meandering SF ≤ 6.2 and F > 0.21.

The existence of meandering flows was recently highlighted by Camnasio *et al.* (2012) and Peltier *et al.* (2014a). This flow is characterized by spatial and temporal periodical oscillations of the central jet from the beginning to the end of the reservoir. The meandering of the jet is the hallmark of the presence of large-scale vortices in the flow, which transfer momentum from the jet towards the rest of the reservoir and induce changes in the velocity distribution, as well as in the sediment transport, compared to a configuration without meandering jet (Peltier *et al.*, 2013). The wave-length of the meander is proportional to the width of the reservoir and/or to the friction coefficient, while the lateral spreading of the jet is affected by the shallowness of the flow, *i.e.* it decreases with increasing friction number, $S = f\Delta B/8H$ (*f* the Darcy-Weisbach friction coefficient) (Chu *et al.*, 1983).

Although the physics of the symmetric and asymmetric configurations are now relatively well understood (Mullin *et al.*, 2003, Canbazoglu and Bozkir, 2004, Kantoush *et al.*, 2008, Dufresne *et al.*, 2010, Camnasio *et al.*, 2011) and the flow modellers have successfully demonstrated that these flow features can be reproduced using an operational numerical model, where only the friction modelling is adjustable (Dewals *et al.*, 2008, Dufresne *et al.*, 2011,

Peng et al., 2011, Camnasio et al., 2013, Khan et al., 2013), many questions remain regarding meandering flows.

In the present article, two meandering flows are experimentally and numerically investigated. The experimental setup is first described. Information is then given on the operational numerical model WOLF2D and the k-ε model we used. A Proper Orthogonal Decomposition (POD) is used for describing the transient behaviour of the jet in the description. The POD analysis enables the comparison of spatial, temporal and energetic characteristics of the experimental and numerical modelling. The results are finally discussed.

2 MATERIAL AND METHOD

2.1 Experimental set-up

The experiments were performed without sediments in a flume located in the hydraulic laboratory of the research group HECE of the University of Liège (ULg), Belgium. The experimental device consists of a horizontal glass channel of 10.40 m long and 0.98 m wide, in which movable blocks were placed to build the rectangular reservoir (Fig. 1).

For these experiments, the inlet channel was 2.00 m long and 0.08 m wide. The reservoir length was set to 1.00 m and the width to 0.98 m. The outlet channel was 1.50 long with the same width as the inlet channel and a tailgate ends the flume. All the surfaces were made of glass, except for the bottom of the flume (polyvinyl chloride) and the converging section (metallic sheets).

The discharge, Q, was measured with an electromagnetic flow-meter (uncertainty of 0.025 L/s). The mean water depth, H, in the reservoir was measured using an ultrasonic probe and the uncertainty was estimated to be equal to 1% of the mean value. The flow dynamics was video-recorded using a high-resolution camera at a rate of 25 Hz during 6′30″ and the surface velocity fields were calculated by LSPIV (Hauet et al., 2008, Kantoush and Schleiss, 2009). The uncertainty on the mean surface velocity was estimated to 5%.

In the present paper, x, y and z denote the longitudinal, the lateral and the vertical directions respectively of the Cartesian reference frame attached to the flume; $x = 0$ immediately downstream from the inlet channel and $y = 0$ at the right bank of the reservoir. $z = 0$ at the bottom of the reservoir.

Two discharges were used, resulting in two different mean water depths in the reservoir. The corresponding Froude number, F, friction number, S, and Reynolds number, $R = U_{in}D/\nu$ (D the hydraulic diameter of the inlet channel and ν the kinematic viscosity of the water at 20°C), in the inlet channels are summarized in Table 1. The Froude numbers of both experiments are close to 0.4 and are a direct consequence of the geometry of the reservoir

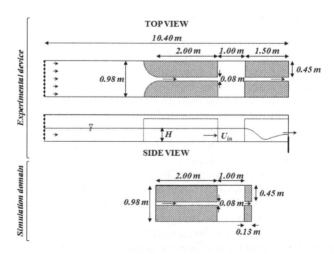

Figure 1. Sketch and pictures of the experimental device and the simulation domain.

Table 1. Main characteristics of the measured flows.

Test ID	Q (L/s)	H (cm)	F	S	R
F	0.50	1.80	0.41	0.10	8,456
NF	1.00	4.20	0.46	0.03	24,267

and of the downstream boundary condition. The friction numbers are characteristic of two different friction regimes. Referring to the work of Chu *et al.* (2004), the flow-case F belongs to the frictional regime (the turbulence scale is mainly driven by the water depth), while the flow-case NF belong to the non-frictional regime (the turbulence scale is mainly driven by the horizontal length-scale). The Reynolds numbers in both cases confirm that the flows are turbulent, but they are hydro-dynamically smooth.

2.2 *Numerical model: WOLF2D*

The numerical modelling of the flows was performed using the flow model WOLF2D developed at the University of Liege (Dewals *et al.*, 2008). It is an operational code used for engineering studies, as well as research applications.

The space discretization of the 2D-H shallow water equations was performed by means of a Finite Volume scheme. The variable reconstruction at cells interfaces was linear and was combined with a slope limiter, leading to second-order space accuracy. The convective fluxes were computed by a Flux Vector Splitting (FVS) method developed by Erpicum *et al.* (2010), which requires low computational cost, is completely Froude-independent and facilitates a satisfactory balance with the discretization of the bottom slope term.

The time integration was performed by means of a 3-step third-order accurate Runge-Kutta algorithm, limiting the numerical dissipation in time that smears the dynamics of the vortex generation in the jet. For stability reasons, the time step was constrained by a Courant-Friedrichs-Lewy condition based on gravity waves and set to 0.2.

The Darcy-Weisbach standard formulation was used for the friction modelling (Dufresne *et al.*, 2011) and the bottom roughness was set to $k_s = 0.1$ mm (representative of the polyvinyl-chloride state during the experiments).

A depth-averaged k-ε model with two different length-scales accounting for vertical and horizontal turbulence mixing was used (Erpicum *et al.*, 2009). The horizontal turbulence mixing was classically modelled using the two additional transport equations k and ε, while the vertical turbulence mixing was treated with an algebraic model (Fisher Model: $v_t = 0.08 H u^*$, v_t the turbulent eddy viscosity and u^* the friction velocity).

In the simulations, the inlet channel of 2 m was reproduced in order to have an injection in the reservoir as close as possible to the experimental configuration (Fig. 1). The experimental discharges were used as inflow conditions at the beginning of the inlet channel, but the unit discharge was not uniformly distributed. As prescribed by Dewals *et al.* (2008), the unit discharge distribution must be slightly disturbed for introducing an asymmetry in the flow, which will generate the oscillation of the jet. The disturbance linearly varied from −1% to 1% across the inflow cells. The water depth measured 13 cm downstream from the outlet of the reservoir was prescribed as downstream boundary condition.

The simulation was performed in two steps. (i) a periodic regime was reached after a run of 3000 s. (ii) The last time step of the first simulation was then used as initial and boundary conditions for the second simulation, which was recorded during 500 s at 25 Hz in order to be compared with experiments.

2.3 *Proper orthogonal decomposition (Holmes et al., 2012)*

The meandering of the jet is characterised by spatial and temporal periodical oscillations (Fig. 2), which are the hallmarks of the contribution of different types of coherent structures

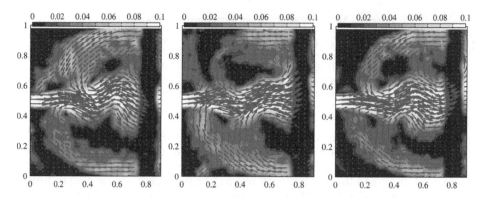

Figure 2. Experimental instantaneous velocity fields of flow-case F. The black area around $x = 0.8$ m is a blank zone in the measurements.

convected in the flow (Peltier *et al.*, 2014a, Peltier *et al.*, 2014b). Given the unsteady nature of the flows, the classical parameters (velocity fields, Reynolds shear stresses) usually used for describing the flows are not sufficient here for comparing experiments and numerical modelling. A method is needed for describing the unsteady characteristics of the flow.

Considering a collection of N snapshots of the fluctuating horizontal velocity fields, $\underline{u}'(x_p, y_p, t_n) = \underline{u}'(\underline{x}, t)$ ($n, p \in \mathbb{N}^*$, $\Delta t = t_{n+1} - t_n = CST$ and $\underline{x} \in \Omega \subset \mathbb{R}^2$), which are square integrable functions (*i.e.* $\underline{u}'(\underline{x}, t) \in L^2(\Omega)$), the Proper Orthogonal Decomposition (POD) provides an orthonormal basis of M spatial functions $\underline{\phi}_m(\underline{x})$ of $L^2(\Omega)$, called spatial modes, and an orthogonal basis of M temporal coefficients, $a_m(t)$ ($m \in \{1, ..., M \leq N\}$ and $M, N \in \mathbb{N}^*$), such that:

$$\min\left(\frac{1}{N} \sum_{n=1}^{N} \left\| \underline{u}'(\underline{x}, t_n) - \sum_{m=1}^{M \leq N} a_m(t_n) \underline{\phi}_m(\underline{x}) \right\|_{L^2}^2 \right) \tag{1}$$

$\| \ \|_{L^2}$ being the induced norm in $L^2(\Omega)$ (*i.e.* the root mean square of the inner product for $L^2(\Omega)$).

The POD proceeds in three steps.

1. The temporal correlation matrix C is first calculated:

$$C_{ij} = \sum_{p=1}^{P} \underline{u}'(\underline{x}_p, t_i) W_{kk} \underline{u}'(\underline{x}_p, t_j) \quad C \in \mathbb{R}^{N \times N}, W \in \mathbb{R}^{P \times P} \tag{2}$$

W is a diagonal weighting matrix, for which the elements along the diagonal are the cell volumes of each of the P grid points of one snapshot.

2. The temporal coefficients $a_m(t)$ are secondly deduced from the resolution of the following eigenvalue problem with the *eig* function of Matlab:

$$\frac{1}{N} \sum_{j=1}^{N} C_{ij} \alpha_m(t_j) = \lambda_m \alpha_m(t_i) \tag{3}$$

C is definite, positive and symmetric. As a consequence, the eigenvalues λ_m are all real with $\lambda_1 \geq \lambda_2 \geq \cdots \geq \lambda_N > 0$, and the eigenvectors $\alpha_m(t)$ are orthonormal. The temporal coefficients, $a_m(t)$, are function of the eigenvectors and of the eigenvalues and they are orthogonal:

$$a_m(t) = \sqrt{N \lambda_m} \alpha_m(t) \text{ with } \langle a_n \rangle_N = 0 \text{ and } \langle a_n a_m \rangle_N = \lambda_n \delta_{nm} \tag{4}$$

24

3. The spatial modes are thirdly computed by projecting the fluctuating velocity ensemble onto the temporal coefficients, i.e.:

$$\underline{\phi}_m(\underline{x}) = \frac{1}{N\lambda_m}\sum_{n=1}^{N}\underline{u}'(\underline{x},t_n)a_m(t_n) \text{ with } \left\|\underline{\phi}_m\right\|_{L^2}^2 = \underline{\phi}_m^T W \underline{\phi}_m = 1 \tag{5}$$

The spatial modes are orthonormal with respect to the inner product in L^2, $\underline{\phi}_m^T W \underline{\phi}_m$.

3   RESULTS AND DISCUSSION

The POD of the experimental and numerical fluctuating velocity fields allows us to describe unsteady flows by analysing the behaviour of the first spatial modes and temporal coefficients, which are representative of the most energetic coherent structures. The POD was applied on 9,000 fields/snapshots of each measurement and numerical simulation of the flow-cases presented in Table 1.

3.1   *Energy*

By construction of the POD basis, the mth eigenvalues calculated with Equation (3) correspond to the mean fluctuating kinetic energy per unit mass captured by the mth mode, E_m, and the sum of the N non-zero eigenvalues is equal to the mean total fluctuating kinetic energy per unit mass, E_T. The mean fluctuating kinetic energy contained in the mth modes is plotted in Figure 3 and is compared with the simulations. For both cases, the numerical model is able to capture the energy of the first and second modes. For the frictional case F, the third and the fourth modes are paired in the simulation while it is not the case in the experiments, but the energy levels are still in good agreement. From the fifth modes the numerical model underestimated the energy. For the non-frictional case NF, the level of energy from the third mode to the sixth mode of the simulation is similar to the experiment. From the seventh the simulation fails to reproduce the energy distribution and very quickly underestimates the energy level, which indicates that the small random structures with low energy cannot be well rendered by the model.

The computation of the total mean fluctuating kinetic energy indicates that for the frictional case, E_{Tnum} is equal to 86% of E_{Texp}, while for the non-frictional case E_{Tnum} is only equal to 49% of E_{Texp}. This indicates that the model well captured the essential structures for representing the energy in the F-case, while for the NF-case, some essential structures are missing. This is due to the energy drop between modes 2 and 3 in the F-case compared to the NF-case.

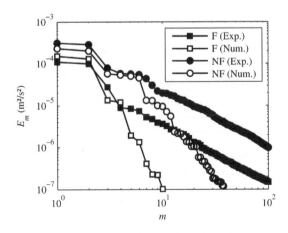

Figure 3.   Mean fluctuating kinetic energy contained in the mth mode.

3.2 Temporal coefficients

The temporal coefficients were calculated with Equation (4) and they are represented for the four first modes for both flow-cases in Figure 4. For the frictional flow case, mode 1 and mode 2 of the experiment have the same patterns and are phase-shifted in time, which indicates that both modes are related to the same structures (Rempfer and Fasel, 1994). The numerical model well represents this behaviour. When considering the third and the fourth modes, no pairing is observed for experiments, while a clear pairing is obtained with the numerical model.

For the non-frictional case NF, a pairing is observed for the four first modes of the experiment and simulation. The complex patterns of the temporal coefficients are well represented for the two first modes. For modes 3 and 4, the simulated patterns are simpler, but numerical and experimental frequencies are close.

The numerical model cannot represent modes greater than 4 for F-case, modes greater than 6 for NF-case. Linking the temporal coefficients with the energy indicates that modes with an energy E_m smaller than 1×10^{-5} m²/s² cannot be well rendered by the numerical model. These low energy modes are parasite motions due to measurement errors or high frequency motions. They have low incidence on the main flow feature (Peltier *et al.*, 2014b).

3.3 Spatial modes

The vorticity of the spatial modes calculated with Equation (5) were used for describing the rotating nature of the meandering flows. Seven transversal profiles of vorticity are plotted in Figure 5 for each flow case (experiments and simulations) and the longitudinal distribution of the vorticity long the reservoir centreline is displayed in Figure 6.

For the frictional-case the numerical model is able to reproduce the lateral distribution of the vorticity of the two first modes until 40 cm downstream from the reservoir inlet. Downstream from this point, the numerical model fails at representing the distribution; some extreme values are indeed simulated. Nevertheless, the meander oscillations along the centreline reservoir are well represented. Concerning mode 3 and mode 4, the model completely fails at representing the spatial modes.

The numerical model captures better the physics of the non-frictional flow. Except at some locations for the third mode, the model and the experiments are in good agreement. The computed lateral distribution of vorticity as well as the longitudinal one are in a good agreement with the experiments.

The previous observations are consistent with the findings obtained for the temporal coefficients. When a spatial mode corresponds to a mode with low energy, the numerical model fails to reproduce it accurately. As the non-frictional flow is more energetic than the

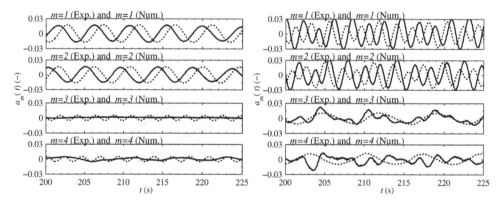

Figure 4. Temporal coefficients of the fourth first modes of the POD analysis. Plain lines correspond to experiments and dotted lines to simulations. (left figure) flow-case F. (right figure) flow-case NF.

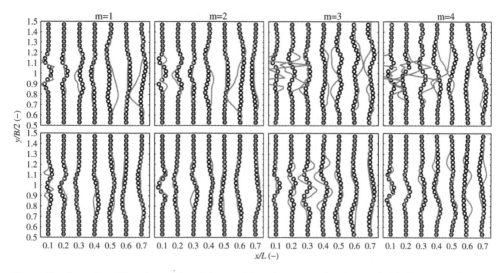

Figure 5. Lateral profiles of vorticity of the spatial modes; o experiments, — simulations. (Top figures) F-case. (Bottom figures) NF-case.

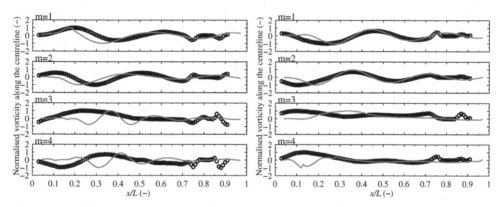

Figure 6. Longitudinal profile of vorticity of the fourth first spatial modes; o experiments, — simulations. (Left figure) F-case. (Right figures) NF-case.

frictional flow, and has more modes with energy greater than 1×10^{-5} m²/s², the overall characteristics of the non-frictional flow-case are better reproduced.

4 CONCLUSION

The present paper investigates experimentally and numerically two meandering flows in shallow rectangular reservoir. In the experiments, the reservoir boundary conditions were chosen to obtain flows with different friction regimes (frictional and non-frictional, see Chu *et al.* (2004)) and the flow dynamics was quantified by LSPIV. The flows were then modelled using WOLF2D, which solves the shallow water equations and uses a depth-averaged k-ε turbulence model. The numerical boundary conditions were taken as close as possible to the experiments.

A Proper Orthogonal Decomposition was performed on the measured and simulated fluctuating velocity fields for describing the main flow features. The POD indeed discriminates the structures in the flow in terms of their relative contributions to the mean total fluctuating kinetic energy and enables the description of unsteady flows.

The numerical model is able to represent POD modes, which have a mean fluctuating kinetic energy at least greater than 1×10^{-5} m²/s², whatever the friction regime of the modelled flow. This corresponds generally to the two to four first modes of the POD.

ACKNOWLEDGMENT

The research was funded by the University of Liège (grant SFRD-12/27). The authors are grateful for the assistance provided by the research technicians during the experiments and the fruitful discussions about the POD analysis with Prof. Vincent Denoël.

REFERENCES

Camnasio, E., Erpicum, S., Orsi, E., Pirotton, M., Schleiss, A.J. & Dewals, B. 2013. Coupling between flow and sediment deposition in rectangular shallow reservoirs. *Journal of Hydraulic Research.*

Camnasio, E., Orsi, E. & Schleiss, A.J. 2011. Experimental study of velocity fields in rectangular shallow reservoirs. *Journal of Hydraulic Research,* 49, 352–358.

Camnasio, E., Pirotton, M., Erpicum, S. & Dewals, B. 2012. Experimental and numerical investigation of a meandering jet in shallow rectangular reservoirs under different hydraulic conditions. 3rd International Symposium on Shallow Flows. Iowa City, USA.

Canbazoglu, S. & Bozkir, O. 2004. Analysis of pressure distribution of turbulent asymmetric flow in a flat duct symmetric sudden expansion with small aspect ratio. *Fluid Dynamics Research,* 35, 341–355.

Chu, V.H., Khayat, R.E. & Wu, J.H. 1983. Stability of turbulent shear flows in shallow channel. 20th IAHR Congr., Sep. 5–9, Moscow, USSR. Delft, The Netherlands, IAHR.

Chu, V.H., Liu, F. & Altai, W. 2004. Friction and confinement effects on a shallow recirculating flow. *Journal of Environmental Engineering and Science,* 3, 463–475.

Dewals, B.J., Kantoush, S.A., Erpicum, S., Pirotton, M. & Schleiss, A.J. 2008. Experimental and numerical analysis of flow instabilities in rectangular shallow basins. *Environmental Fluid Mechanics,* 8, 31–54.

Dufresne, M., Dewals, B.J., Erpicum, S., Archambeau, P. & Pirotton, M. 2010. Classification of flow patterns in rectangular shallow reservoirs. *Journal of Hydraulic Research,* 48, 197–204.

Dufresne, M., Dewals, B.J., Erpicum, S., Archambeau, P. & Pirotton, M. 2011. Numerical investigation of flow patterns in rectangular shallow reservoirs. *Engineering Applications of Computational Fluid Mechanics,* 5, 247–258.

Erpicum, S., Dewals, B.J., Archambeau, P. & Pirotton, M. 2010. Dam break flow computation based on an efficient flux vector splitting. *Journal of Computational and Applied Mathematics,* 234, 2143–2151.

Erpicum, S., Meile, T., Dewals, B.J., Pirotton, M. & Schleiss, A.J. 2009. 2D numerical flow modeling in a macro-rough channel. *International Journal for Numerical Methods in Fluids,* 61, 1227–1246.

Hauet, A., Kruger, A., Krajewski, W.F., Bradley, A., Muste, M., Creutin, J.D. & Wilson, M. 2008. Experimental system for real-time discharge estimation using an image-based method. *Journal of Hydrologic Engineering,* 13, 105–110.

Holmes, P., Lumley, J.L., Berkooz, G. & Rowley, C.W. 2012. Turbulence, Coherent Structures, Dynamical Systems and Symmetry. Second Edition, Cambridge University Press.

Kantoush, S.A., De Cesare, G., Boillat, J.L. & Schleiss, A.J. 2008. Flow field investigation in a rectangular shallow reservoir using UVP, LSPIV and numerical modelling. *Flow Measurement and Instrumentation,* 19, 139–144.

Kantoush, S.A. & Schleiss, A.J. 2009. Large-scale piv surface flow measurements in shallow basins with different geometries. *Journal of Visualization,* 12, 361–373.

Khan, S., Melville, B.W., Shamseldin, A.Y. & Fischer, C. 2013. Investigation of flow patterns in storm water retention ponds using CFD. *Journal of Environmental Engineering (United States),* 139, 61–69.

Mullin, T., Shipton, S. & Tavener, S.J. 2003. Flow in a symmetric channel with an expanded section. *Fluid Dynamics Research,* 33, 433–452.

Peltier, Y., Erpicum, S., Archambeau, P., Pirotton, M. & Dewals, B. 2013. Experimental and numerical investigation of meandering jets in shallow reservoir: potential impacts on deposit patterns. In Nguyen, K.D., Benoit, M., Guillou, S., Sheibani, N., Philipps, J.G. & Pham Van Bang, D. (Eds.) THESIS 2013, Two-pHase modElling for Sediment dynamIcs in Geophysical Flows. Chatou, France, SHF—EDF R&D.

Peltier, Y., Erpicum, S., Archambeau, P., Pirotton, M. & Dewals, B. 2014a. Experimental investigation of meandering jets in shallow reservoir. *Environmental Fluid Mechanics*.

Peltier, Y., Erpicum, S., Archambeau, P., Pirotton, M. & Dewals, B. 2014b. Meandering jets in shallow rectangular reservoirs: POD analysis and identification of coherent structures. *Accepted in Experiments in Fluids*.

Peng, Y., Zhou, J.G. & Burrows, R. 2011. Modeling free-surface flow in rectangular shallow basins by using lattice boltzmann method. *Journal of Hydraulic Engineering*, 137, 1680–1685.

Rempfer, D. & Fasel, H.F. 1994. Evolution of three-dimensional coherent structures in a flat-plate boundary layer. *Journal of Fluid Mechanics*, 260, 351–375.

Reservoir Sedimentation – Schleiss et al. (Eds)
© *2014 Taylor & Francis Group, London, ISBN 978-1-138-02675-9*

Combination of 2D shallow water and full 3D numerical modeling for sediment transport in reservoirs and basins

G. Wei & M. Grünzner
Flow Science Inc., Santa Fe, NM, USA

F. Semler
Flow Science Deutschland GmbH, Rottenburg am Neckar, Germany

ABSTRACT: This article is dealing with the combination of two different numerical model types—the 2D shallow water model and the 3D model for the simulation of sediment transport. For the simulation of the sedimentation and erosion processes in reservoirs and basins it is a challenge to model all in one full model. Due to the size of catchment and storage areas/volumes of the reservoirs and the necessary high resolution near the barrier, it is in most real cases impossible to set up one full three dimensional numerical model to simulate the processes of sediment transport. The sediment transport at reservoirs is on the one hand the deposition of the transported material from the inflow of a river and, on the other hand, the erosion process during a sediment flushing event. If sediment transport processes should be investigated on the catchment scale, it is necessary to simulate the erosion and deposition (transport) process in the river itself to get the information of the sediment influx in the reservoir related to the hydrograph. This information is necessary if, for example, a turbidity current should be directly flushed through the reservoir during a natural flood event. State of the art is to model it in parallel. The processes in the river, on the catchment scale, with a 2D shallow water model and the flushing at the barrage in an extra 3D model to resolve the vertical flow and erosion behavior. Sometimes a third, higher resolved 2D model is needed to model the hydraulics in the reservoir higher resolved than the river in the catchment area. This paper presents the advantage of the direct combination by coupling of 2D shallow water approach for large areas and the full 3D Navier-Stokes simulation for the volumes with vertical influence. With this method it is possible to run a full flood scenario on catchment area scale including the barrage without the necessity of a model separation and subsequent modeling of boundary conditions. This approach helps to get a better understanding of the sediment transport processes in reservoirs and basins.

1 INTRODUCTION

Generally in sediment transport simulations one has to separate suspended load and bed-load simulation on one hand and modeled dimensions (2D or 3D) on the other hand. The separation between suspended and bed-load transport is founded in the fact that the different transport behavior (entrained in the fluid or rolling and saltation on the bed) is not consistently describable by only one model. This is the reason for the strict separation of the different transported material. Also on the model side it has been necessary to separate between the simulated dimensions. On one side some problems, especially those on catchment area scale, are limited by hardware resources and on the other side the vertical flow could be neglected. However, if there is an area with significant vertical flow or non-hydrostatic pressure in a 2D model, a detached 3D model is necessary. With FLOW-3D it has been possible to run a shallow water model or a full 3D one, furthermore a sediment investigation with suspended and bed-load material by solving the momentum and density equation respectively a transport equation. With the new FLOW-3D version 11 (scheduled release 2014) it is now possible to combine all elements and

Table 1. Designated model combinations in FLOW-3D v11.

	suspension	bedload	both
2D	✚	✚	✚
3D	✚	✚	✚
hybrid 2D + 3D	✚	✚	✚
multi block	✚	✚	✚

run a coupled simulation. On a hybrid mesh, comprising of shallow water and 3D mesh blocks, a combined (suspended and bed-load) sediment transport simulation can be run. Any possible combination of the individual modules is possible as well as any combination of the different mesh types. Also the "multi-block" logic is available which gives the possibility to nest multiple mesh blocks into each other or to place a 3D block into a shallow water block for a higher local resolution.

2 SIMULATION TECHNIQUE

2.1 *Shallow water model*

The shallow water model in FLOW-3D can consist, analog to the 3D model, of multiple mesh blocks. The mesh blocks must comprise of two vertical cells. It is important to arrange the mesh in a way, that all the geometry and all the water is mapped in the bottom cell and the cell is not filled more than 90% (component and water). This is easy to achieve by defining the appropriate cell height of the bottom mesh layer. Another requirement is the definition of gravity in negative z-coordinate. Only the horizontal velocities in x- and y-directions are solved. The calculation of the pressure is reduced to a hydrostatic profile. To calculate the bottom shear stress the following approach is used.

$$\tau = \rho \cdot C_D |\bar{u}| \bar{u} \tag{1}$$

The user can choose between a constant drag coefficient C_D (default is 0.0026) or a calculated one. The calculated C_D is a relation of the flow depth and surface roughness. Instead of surface roughness, the sediment describing diameter or if multiple fractions are defined, the calculated d_{50}, could be used for the C_D calculation (FLOW-3D 2014a).

2.2 *3D model*

Block structured mesh blocks are used in the 3D part of the model as well. The mesh blocks can be arranged arbitrarily. If a mesh block is placed into another mesh block for a higher local resolution, this block is called a "nested block". It is now possible to combine a 3D mesh block with a shallow water mesh block by nesting the 3D into the shallow water one, for a local higher and three dimensional resolution (hybrid mesh block). Figure 1 shows such hybrid shallow water and 3D mesh block combination. Figure 2 shows, that it is possible to model a whole reservoir dam in one 3D model.

2.3 *Entrainment*

The fractions defined as suspended material (entrained), are transported by the fluid like in solution. The suspension movement is calculated by solving the Navier-Stokes equations. The density of the fluid is evaluated in each grid cell taking into account the densities of the water and of the suspended materials. In this way it is possible to model turbidity currents. To define a sediment fraction as suspended material the bed-load coefficient must be set to zero. Technical details are described in Wei et al. (2014) and Brethour and Burnham (2010).

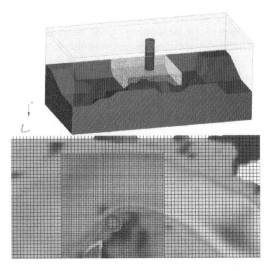

Figure 1. Example of a nested 3D mesh block located in a shallow water mesh block for a higher resolution of the flow around the pier. The horizontal cut shows the grid lines.

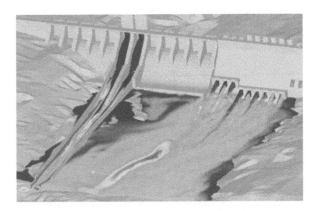

Figure 2. Example of a 3D hydraulic simulation of a dam in FLOW-3D (FLOW-3D 2014b).

2.4 *Bed-load model*

The implemented bed-load function is the threshold approach of Meyer-Peter and Müller (1949) and is described in Brethour and Burnham (2010) as well as in Wei et al. (2014). It is possible to define multiple species in the setup. A transport equation is solved for every fraction. Because of the spatial variation of the composition of the bed-load material the characteristic sediment diameter (d_{50}), which characterizes the bed roughness, is recalculated in every grid cell and each time step. In order to account for different bed shapes a roughness multiplier can be defined by the user. Although multiple fractions are possible, FLOW-3D version 11 won't be able to capture the armoring effect. This is due to the assumption that the material is distributed homogeneously in one simulation cell.

2.5 *Total load—the combination of entrainment and bed-load*

Suspended load and bed-load can be simulated parallel in one model. Multiple fractions can be defined, which are transported individually. Regarding their material properties and impact of the shear stress, they are able to erode, move or deposit individually. For each sediment fraction it is possible to specify individual transport parameters or transport functions.

33

This is especially important for materials in transition between suspended or bed-load transport, since the transport form is also depending on the hydraulic load. Furthermore an individual threshold for the dimensional critical Shields number can be defined for each fraction. The user can alternatively use the approach of Shields-Rouse (FLOW-3D 2014a) and let the code calculate the actual threshold. Further technical details are in Brethour and Burnham (2010) as well as in Wei et al. (2014).

3 INVESTIGATIONS

An adequate test case, which can be set up with limited consumption of resources, is a scouring process. A horizontal submerged jet is eroding (after an immobile sediment bed) a scour hole and deposits the material adjacently. This is well documented for example in Chatterjee et al. (1994). The setup used in the physical model test from Chatterjee et al. (1994) is shown in Figure 3.

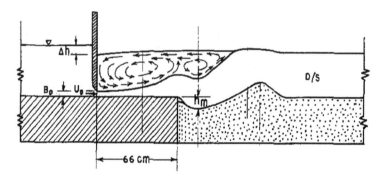

Figure 3.   Physical model test from Chatterjee et al. (1994), investigated 2D and 3D with FLOW-3D.

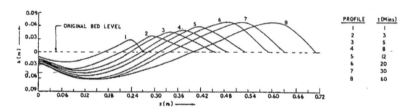

Figure 4.   Development of the scour Chatterjee et al. (1994).

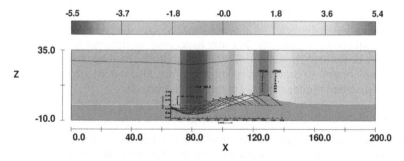

Figure 5.   Comparison of the erosion shape. The contour plot gives the FLOW-3D results of the change of the packed bed elevation and the lines are the results of Chatterjee et al. (1994).

34

The physical model test was installed in a 60 centimeter wide flume. The published results are related to the symmetry axis of the flume for which the wall influence can be neglected. Therefore it is permissible to take a 2D approach and model just a vertical mesh slice. In this context, two dimensional doesn't mean depth averaged or shallow water but considering the flow in a x, z plane neglecting a variation of the flow variables normal to this plane.

Figure 5 shows the good accordance of the calculated scour shape to the measured one.

The next example shows a simplified three dimensional erosion process downstream a hydraulic jump blown out of a stilling basin.

The picture in Figure 6 was created with the newly implemented post processor FlowSight™, which allows to display multiple iso-surfaces in one plot.

Also for the shallow water model the erosions can be plotted in 3D. The following Figure 7 presents a shallow water test case. The original topography (upper left) is deformed by the

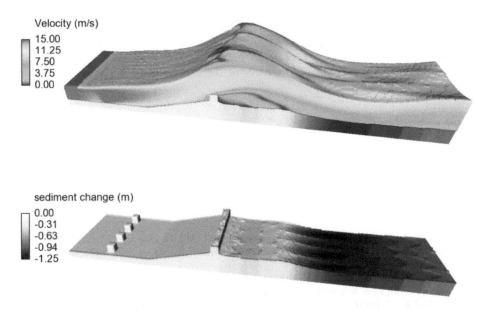

Figure 6.   Erosion development downstream a hydraulic jump blown out of a stilling basin.

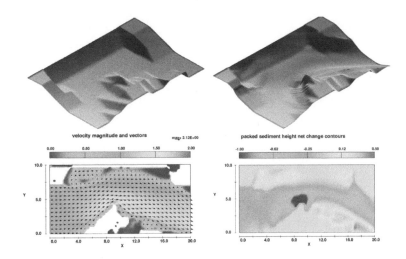

Figure 7.   Erosive deformation of the terrain. Original topography (upper left) deformed state (upper right). Flow velocity with vectors (lower left) and net bed change in meter (lower right).

erosion and deposition (upper right). It is nice to see in that figure, that the areas of backflow or stagnation are trending to sand-up and the main channel trends to erode.

4 CONCLUSIONS

These first and simple test cases on this very interesting topic illustrate the enormous advantage of the model combination for hydraulic engineering. Especially the combination of 2D shallow water and full 3D modelling enabling to model an entire catchment area including the reservoir and the barrier with all significant hydraulic characteristics. With this capability it is no longer necessary to have internal model switches with all the limitations of defining boundary conditions and making assumptions on the model interconnections. Especially the definition of transport functions respectively the generation of discharge hydrographs on these interfaces is now history. Even the boundary effects, which always happen on boundary conditions, are now only appearing on the global outer boundary conditions.

Since we haven't yet an appropriate data set for a model validation on catchment area scale, or an entire reservoir (e.g. flushing event) we want to place here a call for it. If you have access to an appropriate data set, please contact one of the authors listed in the article head. We appreciate that in advance.

REFERENCES

Brethour, J., Burnham, J. (2010) Modelling sediment erosion & deposition with FLOW-3D. *Technical Note 85, Flow Science Inc.*, Santa Fe, NM, USA.
Chatterjee, S., Ghosh, S., and Chatterjee, M. (1994). "Local Scour due to Submerged Horizontal Jet." *J. Hydraul. Eng.*, 120(8), 973–992.
FLOW-3D (2014a) Software Documentation, FLOW-3D Version 11. *Flow Science Inc.*, Santa Fe, NM, USA.
FLOW-3D (2014b) Training Material, FLOW-3D Version 11. *Flow Science Inc.*, Santa Fe, NM, USA.
Meyer-Peter, E, Müller, R. (1949) Eine Formel zur Berechnung des Geschiebetriebs. Schweizerische Bauzeitung, 67. Jahrgang.
Wei, G., Brethour, J., Gruenzner, M., Burnham, J. (2014) The sediment scour model in FLOW-3D, Technical Note 99, *Flow Science Inc.*, Santa Fe, NM, USA.
Note: **FLOW-3D** and **TruVOF** are registered trademarks of Flow Science, Inc. in the USA and other countries.

Reservoir Sedimentation – Schleiss et al. (Eds)
© *2014 Taylor & Francis Group, London, ISBN 978-1-138-02675-9*

Photometric analysis of the effect of substrates and obstacles on unconfined turbidity current flow propagation

J.M. McArthur, R.I. Wilson & H. Friedrich
University of Auckland, Auckland, New Zealand

ABSTRACT: Experimental flows of turbidity currents are investigated using photometric analysis. The present work studies the interaction of turbidity currents with different obstacles and substrates, using a 25 mm square-bottomed cylinder obstacle, and a 0.8 mm rough surface for the entire base of the test basin. Notable differences are evident in the nature and characteristics of current flows. The rough substrate decreases current velocity and minimises the appearance of turbidity current phenomena, such as lobe and cleft formation and Kelvin-Helmholtz billows. Presence of an obstacle causes localised decreases in velocity, but otherwise has little effect on overall velocity of the flow. Lobe and cleft formations increase significantly after the current passes over an obstacle. It is concluded that these findings warrant the inclusion of rough substrates in further experimental testing. A generally used smooth laboratory substrate does not take into account substrates encountered in nature, and our preliminary study shows that there are significant flow characteristic differences. The presented work is based on limited tests, it is recommended to undertake a more comprehensive study to evaluate the substrate roughness effect.

1 INTRODUCTION

Turbidity currents are a type of gravity flow where sediment-laden fluid flows through deep ocean or lake environments, driven by a higher relative density of the fluid to the ambient water (Simpson, 1982). They can cause substantial damage to submarine structures (Ermanyuk & Gavrilov, 2005), including oil well caps and oil or gas pipelines (Gonzalez-Juez et al. 2009). Limited research exists into the nature of these flows in regard to their interactions with obstacles in an unconfined environment. Nogueira et al. (2012) did study the influence of bed roughness in a confined environment. Based on the previous studies, the hypothesis was developed that by using photometric analysis of experimental current flows and examining the characteristics of turbidity currents as they interact with obstacles and rough substrates, further insights into the dynamics of turbidity current flow can be obtained.

The study was thus structured as a preliminary comparison of current interactions with different configurations of rough substrates and obstacles. Three obstacle/substrate setups are used:

a. Flow of an unconfined turbidity current over a smooth glass surface.
b. Flow of an unconfined turbidity current over a 25 mm square-bottomed cylinder running perpendicular to flow direction, on a smooth glass surface.
c. Flow of an unconfined turbidity current over a textured substrate of 0.8 mm roughness height.

2 METHODOLOGY

2.1 *Experimental methodology*

Testing of experimental lock-exchange flows was undertaken at the Fluid Mechanics Laboratory at the University of Auckland. The existing purpose-built apparatus for

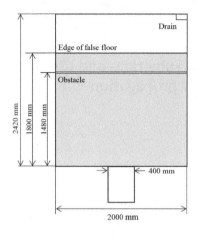

Figure 1. (a) Plan view diagram of basin setup. (b) Fixed rough substrate used for the third test scenario.

turbidity current testing consisted of a basin, length of 2420 mm, width of 2000 mm and height of 600 mm (Fig. 1a), and adjacent lockbox. The basin was constructed on the concrete floor of the laboratory, with 12 mm clear Perspex glass walls and plywood framing. Test flows travelled over a smooth glass false floor at 200 mm height above the concrete floor. A lockbox, situated at the proximal end of the basin, held the sediment-laden fluid before release.

Fluid for the sediment-laden flow was created as a slurry of powdered kaolinite clay and Ballotini (2% each, by volume) in water, which resulted in a current density in the range of 1075–1082 kgm⁻³. Use of this mixture was justified by previous research recommending the proportions as optimal for generating a sufficient density gradient for turbidity current flow in an experimental setting (Sangster, 2011).

The obstacle selected for the second test scenario was a 25 mm × 25 mm square-bottomed aluminium cylinder. The height of 25 mm was chosen to roughly approximate the size and shape of a submarine cable, as these are common obstacles that a turbidity current in nature may interact with.

The final test scenario incorporated the use of a fixed rough substrate, which was attached to the glass false floor (Fig. 1b). The substrate consisted of 0.8 mm diameter sand glued evenly over a 4 mm thick silicon plastic sheet.

Use of multiple recording instruments, including plan and elevation cameras and a digital video camera, meant that points of reference were required to synchronise the images captured, for correct analysis.

To provide this reference between the cameras, the video camera started recording first, and plan view camera was triggered using a remote switch, at the point when the lockbox gate lifted from the floor of the basin. Oral direction from the person operating the plan view camera was given to the person operating the elevation view camera to ensure synchronisation. As images were captured at 4.5 frames per second, a small margin of error was likely between the starting frames for each camera. To adjust for this, profiles created for the plan and elevation view were matched by finding the furthermost point on the plan profile (in the longitudinal direction) and matching it to the elevation profile that had progressed to the same point in the longitudinal direction.

The following steps were undertaken during the experimental testing:

a. Lighting setup rigged into place and switched on, all other lighting in Fluid Mechanics Laboratory switched off (Fig. 2a).
b. Cameras rigged into position, focused and set to appropriate zoom by connecting to laboratory computer and adjusting settings in Camera Pro 2.0.
c. Cameras disconnected from computer, remote switch attached to plan view camera.

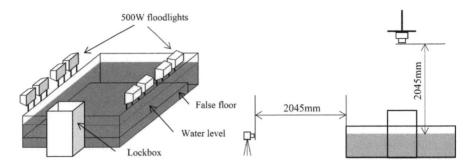

Figure 2. (a) Lighting configuration for optimal illumination of turbidity current flow. (b) Plan and elevation camera setup.

d. Slurry of water, kaolinite and Ballotini mixed in large bucket.
e. Video camera started recording.
f. Slurry poured into lockbox and lockbox gate lifted slowly and evenly.
g. Plan and elevation view cameras begin capturing images at the point where lockbox gate lifts off basin floor.
h. Cameras continue to capture images at 4.5 frames per second until the flow reaches the end of the false floor.

2.2 *Digital camera setup and illumination*

Two Nikon D90 digital cameras captured images of the turbidity current flow in plan and elevation view (Fig. 2b). The plan view camera was attached to a rig, suspended from the ceiling of the laboratory, facing directly downwards. A second camera was set on a tripod at the side of the basin to capture elevation view images, equidistant from the basin wall as the plan view camera was from the water surface.

Camera mode was set to manual so that all settings could be adjusted. Shutter speed was set to 1/200, and aperture 3.5 to obtain optimal image quality. Capture mode was set to 'continuous high speed', with a frame rate of 4.5 frames per second. A video of each test run was recorded using a Casio Lumix camera, mounted approximately 2 m away from the basin walls at the side. Due to the relatively poor video resolution, video data collected was for observational purposes only and not used for analysis.

Eight 500 W floodlights provided illumination to the testing basin (Fig. 2a). Photometric analysis measures the amount of light and dark in an image, thus it was a priority to arrange lights in the best configuration to obtain the best level of clarity and contrast in captured images. To find the best configuration, four different variations were trialed in preliminary tests, and images processed in **MATLAB** to evaluate their suitability.

2.3 *Photometric methodology*

Recently, photometric analysis of gravity current flow is becoming more popular (Nogueira et al. 2013). The photometric analysis processes each captured image, to obtain a profile of the current's outer boundaries for each image, and use these to create flow progression contour plots (plan and side view), to examine the flow and compare the interactions with different substrates and obstacles. Images captured on the Nikon D90 digital camera were saved first on to the camera's memory, and later uploaded to a PC for analysis.

Pre-processing of images in Adobe Photoshop involved the removal of known lens distortion from the Nikon D90 camera, and adjustment of images to optimal contrast and brightness for processing. For each test run, the appropriate lens distortion and rotation correction was set to ensure basin sides were vertical and parallel. Brightness and contrast were also adjusted to improve image quality for processing. These modifications were recorded as an Action in Photoshop, and applied using Batch Process, to modify each image in a photo-series.

Corrected images were processed using photometric analysis in MATLAB to generate current profiles for each frame. Custom code was developed in MATLAB to generate the most accurate and clear results. For each experimental test, the image series contained between 80–110 frames. Steps carried out in MATLAB processing are detailed below; Figures 3 and 4 illustrate the transformations for plan and elevation views, respectively.

The processing code used the following steps to generate the required visualisations (Fig. 5a–c):

a. Read jpeg image file into MATLAB.
b. Crop image to area of interest.
c. Convert to gray scale, black out UDVP rig.
d. Apply threshold to produce binary image.
e. Filter binary image to remove areas of light.
f. Generate profile of object in binary image.

In another study (Wilson & Friedrich, 2014), Ultrasonic Doppler Velocity Profiler (UDVP) measurements were used to better quantify the turbulence characteristics of turbidity currents. Thus, a measurement rig is visible in the plan view contour plots, and head outlines are not available for the sections occupied by the UDVP measurement rig.

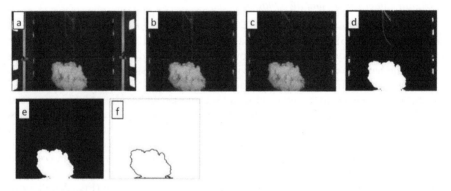

Figure 3.   Plan view image processing.

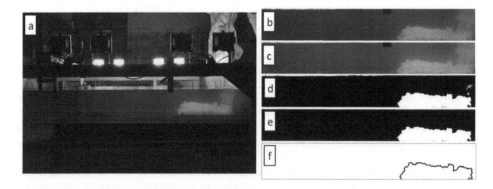

Figure 4.   Side view image processing.

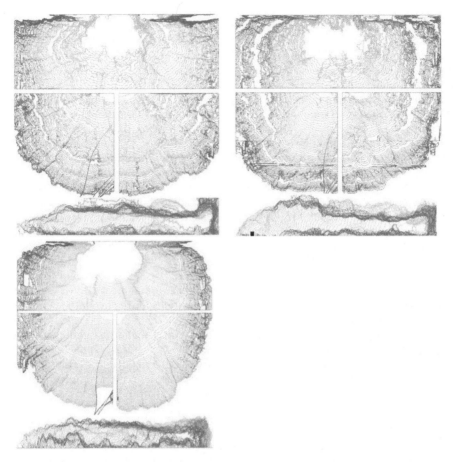

Figure 5. Plan and elevation velocity contour plots: (a) Control scenario. (b) Obstacle scenario. (c) Rough substrate scenario.

3 RESULTS AND DISCUSSIONS

Profiles for each image in a series were plotted to create a contour plot, mapping the progression of the current and using a graduated colour scale to indicate current progression (Fig. 5a–c).

3.1 *Comparison to previous research*

Initial research into turbidity current flow found several conclusions that could be compared with results for this study. Experimental tests over a rough substrate with a variety of heights (0.7 mm–3.0 mm) showed minimal effects in the initial phase of motion, but in the latter phase the velocity reduction is significant (La Rocca et al. 2008). Velocity was measured by the extent of the current head over a set of continuous-capture images from above. Our results, using contour line spacing as an approximation for relative velocities, confirmed La Rocca's et al. (2008) observation. Other research indicated that gravity currents showed a raised current head and decreased velocity leading up to an obstacle (Lane-Serff et al. 1995); this was not evident in testing, however the differences in experimental setup may account for this. Lane-Serff et al. (1995) tested a gravity current over a triangular prism-shaped obstacle, approximately half the height of the current, whereas our study examined a square-bottomed cylinder of a height around 1/10th of the current height. The difference in proportions meant that in our study the obstacle had less impact on the flow.

3.2 Significance of results

Several key insights gained from the analysis have implications for the behaviour of turbidity currents in nature, and important considerations for experimental testing. It needs to be noted that those considerations are based on limited tests, and a more comprehensive study is needed to validate the presented initial observations.

The rough substrate minimised Kelvin-Helmholtz billows, maintaining the shape of lobes formed early in the flow. In minimising the billows, it absorbed kinetic energy in the current and caused the overall current velocity to decrease relative to the control setup. As most turbidity currents in nature flow over an ocean or lake floor, it is realistic to assume that a 0.8 mm rough substrate is more representative of actual conditions than a smooth glass substrate. Based on the marked difference in the current flow over the rough element, the inclusion of rough substrates in experimental testing could be considered as a requirement for results relevant to real-life circumstances. To validate this claim, further testing using a range of roughness heights should be undertaken to fully understand the relationship between substrate roughness and current flow characteristics.

Furthermore, the reduction in development of lobe and cleft formations as the current progressed over a rough substrate also holds implications relating to the real-life occurrence of currents. The formation, growth, and subdivision of lobes in the current form are the predominant method by which the flow progresses forwards and breaks down. As the rough surface slows down this process, there are likely implications for the magnitude of the forces in the flow, and local velocities in the different flow regions.

Results obtained in this study illustrate the importance of the topography a current is traversing in determining the nature and characteristics of the flow. Thus, the importance of incorporating obstacles and roughness in substrates is highlighted for further research, to better mimic the behaviour of turbidity currents as they are found in nature.

3.3 Statistical relevance and limitations to testing

Attaining statistical relevance for qualitative analysis of this nature is difficult; as results are non-numerical and turbidity currents display a large degree of random variation in the flow progression. A large number of samples would be required to eliminate all variability and reach definite and irrevocable conclusions. Testing methods were perfected to find the best setup and procedures for accurate data collection. The lighting configuration in particular required an iterative procedure of setting up four different configurations, capturing photos from a test run and processing these images in MATLAB to find the configuration that gave the best clarity in profiles obtained.

Limitations to experimental accuracy were found in capturing images from both a plan and elevation view. The relatively small depth of 265 mm relative to the basin width of 2000 mm meant that the plan view showed very clear outlines of the turbidity current as it progressed, whereas the elevation view contained some cloudiness, especially in the rear sections of the flow. However, this was a trade-off between the plan and elevation views, as the wide basin allowed the accurate observation of unconfined flows, which would otherwise be prevented in a narrower testing structure.

3.4 Recommendations for future work

Further research in this area of turbidity current interactions with obstacles is supported by the findings of this study. In particular, the rough substrate used is more representative of the ocean or lake bed that turbidity currents would encounter in nature, compared to the smooth glass base. Given the strong influence of the rough substrate in inhibiting certain characteristics of the current flow, further research would benefit by including the roughness in order to gain results that are relevant to turbidity currents in nature. Additionally, including the effect of sediment entrainment in turbidity current flow further adds to the relevance of results as the entrainment of sediment into a current, from the bed it is travelling over, can play a strong role in the nature and evolution of the turbidity current.

4 CONCLUSION

a. The presence of a rough substrate of 0.8 mm height caused unconfined turbidity current flow to slow during the latter phase of flow, relative to the control setup.
b. Flow over a rough substrate of 0.8 mm height showed diminished Kelvin-Helmholtz billows and the absence of a hydraulic jump between the flow head and the rear of the flow.
c. Findings of significant influences of the rough substrate on flow velocity and characteristics indicate the need for inclusion of a rough or textured substrate in further experimental testing, to ensure results are relevant to turbidity currents as they occur in nature.
d. Flow over a 25 mm square-bottomed cylinder obstacle showed localised changes in velocity both preceding the obstacle and after traversing it, however the overall current velocity remained the same as that for the control setup.
e. The presence of a 25 mm square-bottomed cylinder caused the number of lobe and cleft formations at the current's leading edge to increase up to a factor of two.
f. Inclusion of sediment entrainment as an experimental variable is recommended for further research.

ACKNOWLEDGEMENTS

The authors would like to thank Fluid Mechanics Laboratory staff Geoff Kirby and Jim Luo, and research student Mikaela Lewis for support and assistance during this study.

REFERENCES

Ermanyuk, E.V. & Gavrilov, N.V. 2005a. Interaction of an internal gravity current with a submerged circular cylinder. *Journal of Applied Mechanics and Technical Physics* 46(2): 216–223.
Gonzalez-Juez, E., Meiburg, E., & Constantinescu, G. 2009. Gravity currents impinging on bottom-mounted square cylinders: flow fields and associated forces. *Journal of Fluid Mechanics* 631: 65–102.
La Rocca, M.C., Sciortino, G. & Pinzon, A.B. 2008. Experimental and numerical simulation of three-dimensional gravity currents on smooth and rough bottom. *Physics of Fluids* 20: 106603.
Lane-Serff, G.F., Beal, L.M., & Hadfield, T.D. 1995. Gravity current flow over obstacles. *Journal of Fluid Mechanics* 292.
Nogueira, H.I.S., Adduce, C., Alves, E., and Franca, M.J. 2012. The influence of bed roughness on the dynamics of gravity currents. *Proc River Flow 2012*: 357–362.
Nogueira, H.I.S., Adduce, C., Alves, E., and Franca, M.J. 2013. Image analysis technique applied to lock-exchange gravity currents. *Measurement Science and Technology* 24(4).
Sangster, T. 2011. A comparative study of confined and unconfined turbidity flows and their associated deposits. *ME Thesis*. University of Auckland, New Zealand.
Simpson, J.E. 1982. Gravity Currents in the Laboratory, Atmosphere, and Ocean. *Annual Review of Fluid Mechanics* 14(1): 213–234.
Wilson R.I. & Friedrich, H. 2014. Dynamic analysis of the interaction between unconfined turbidity currents and obstacles. Submitted to *9th International Symposium on Ultrasonic Doppler Methods for Fluid Mechanics and Fluid Engineering (ISUD-9)*, Strasbourg, France.

Reservoir Sedimentation – Schleiss et al. (Eds)
© 2014 Taylor & Francis Group, London, ISBN 978-1-138-02675-9

Advances in numerical modeling of reservoir sedimentation

S. Kostic

Computational Science Research Center, San Diego State University, San Diego, USA

ABSTRACT: The integral model of reservoir sedimentation presented herein allows for a continuous progression from a stagnation condition behind the dam to an overflow condition when the water interface in muddy pond rises above the spillway crest. This is a serious computational challenge, which can only be overcome by a high-resolution, non-oscillatory numerical scheme, such as the one employed here. The hydrodynamic response of the reservoir to single-flood events is considered first to demonstrate that the model predicts, for the first time, all stages of the turbidity current propagation and interaction with a dam. Next, the morphodynamic response of the reservoir to a series of intermittent floods is analyzed to capture the co-evolution of the river delta and reservoir bed. The model was found to successfully a) quantify the impact of single-flood events such as e.g. flush floods associated with global worming, and b) predict the cumulative loss of storage capacity.

1   INTRODUCTION

Reservoirs enclosed by dams can be used to supply drinking water and water for irrigation, generate hydroelectric power, provide flood control and support other beneficial uses. Reservoir sedimentation can seriously undermine the reservoir performance and impact the areas upstream and downstream of a dam. Turbidity currents can carry fine sediment tens of kilometers along the reservoir bottom. Eventually, sediments that accumulate in front of outlets can impair the operation of the power or water intakes, clog bottom vents, accelerate abrasion of hydraulic machinery, and even interfere with the flood storage. Deposits associated with deltaic sedimentation can produce flooding upstream of the reservoir, impair navigation, impinge on wildlife and their habitat, inundate powerhouses, bury intakes etc. Dams can also significantly alter sediment delivery to the downstream reaches and adversely affect shoreline ecosystems due to streambed degradation, bank failures, increasing scour at bridges and other hydraulic structures, and reduced amount of sediment reaching coastal plains and deltas. For example, accelerated coastal erosion affecting the Mississippi and Nile deltas is attributed to sediment trapped behind dams more than 1000 km upstream (e.g. Morris & Fan, 1997).

Urbanization-induced land use/land cover changes and global warming could further accelerate the loss of storage capacity in reservoirs and add additional stress to water-stressed regions. Changes in land use/land cover act to reduce vegetation and wood available to stabilize hillslopes and channels, increase sediment loads to streams, and speed up flood flows through the waterways. On the other hand, global warming is increasing the frequency and severity of extreme weather events, including catastrophic floods and storms (UNEP, 2013). The current trends suggest that flush floods, often accompanied by mudslides, are becoming one of the most serious climatic disasters.

Reservoirs have been traditionally designed and operated to continuously trap sediment during a finite lifespan, frequently as short as 50–100 years, which will be terminated by sediment accumulation. As the global demand for water and energy increases, many reservoirs worldwide are approaching their designed life expectancy. Yet, little consideration has been given to replacement of reservoirs or measures that could ensure their sustained long-term use.

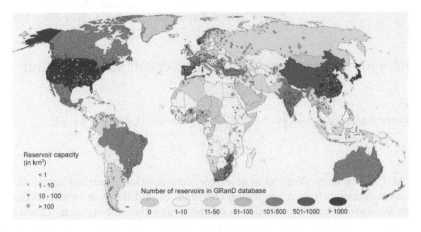

Figure 1. Global distribution of large reservoirs worldwide: GRanD database (Lehner et al., 2011) lists more than 6,862 reservoirs with a cumulative storage capacity of 6,197 km³.

The integral physically-based numerical model of reservoir sedimentation presented here has a potential to become a reliable predictive tool that can be used to evaluate the effect of sediment routing, flushing and other operational procedures in reservoirs, which are potentially effective mitigation strategies. This will become more and more critical in the coming years when the climate-fueled extreme weather patterns are expected to act to additionally reduce the reliability of reservoir flood control and other beneficial uses.

2   NUMERICAL MODEL OF RESERVOIR SEDIMENTATION

The model is based on the model of deltaic sedimentation of Kostic and Parker (2003a, b), which captures the co-evolution of the river-delta morphology and the associated deposits by linking fluvial and turbidity current morphodynamics. The formulation of Kostic and Parker (2003a,b) pertains to turbidity currents that plunge over a steep delta face and continue to run out infinitely in the streamwise direction. That formulation was modified herein in the following way: a) the turbidity current submodel was replaced with the submodel for ponded turbidity currents (e.g. Toniolo et al., 2007) to account for the ponding effect of a dam at the downstream end of the reservoir; b) flood intermittency was taken into account to capture deposits emplaced by realistic floods; c) the outflow boundary condition was selected to reflect that the flow could only vent over an uncontrolled or open-gate controlled spillway; and d) the Ultimate QUICKEST scheme (Leonard, 1991) was replaced with the Kurganov-Tadmore algorithm (Kurganov & Tadmor, 2000) to ensure smooth and continuous computation.

For the sake of consistence, the notation of Kostic and Parker (2003a, b) was implemented throughout this paper as well, such that the equations governing 1D quasi-steady river flow emplacing the fluvial delta take the form:

$$h_f U_f = q_w \tag{1}$$

$$\frac{\partial}{\partial \xi}\left(\frac{1}{2}U_f^2 + gh_f + g\eta_f\right) = -\frac{C_z^{-2}U_f^2}{h_f} \tag{2}$$

$$(1-\lambda_s)\frac{\partial \eta_f}{\partial t} = -I_f\frac{\partial q_S}{\partial \xi} \tag{3}$$

where h_f is the river depth, U_f is the river streamwise velocity, η_f is the elevation of the topset deposit (river bed), ξ is the downslope distance along the fluvial bed, t is time, q_w is the water discharge per unit width, g is the acceleration of gravity, C_z is the Chezy friction coefficient, λ_s is sand porosity, q_s is the volume transport rate of sand per unit width, and I_f is the flow intermittency. The flow intermittency accounts for the fact that the river is morphologically active only I_f fraction of the time.

The equations governing the physics of the 1D unsteady underflow emplacing the bottom-set deposit in a muddy pond behind the dam can be generalized to the following forms:

$$\frac{\partial h}{\partial t} + \frac{\partial Uh}{\partial x} = (1-\delta)e_w U - \delta v_m \tag{4}$$

$$\frac{\partial Uh}{\partial t} + \frac{\partial U^2h}{\partial x} + \delta U v_m = -\frac{1}{2}Rg\frac{\partial Ch^2}{\partial x} - RgCh\frac{\partial \eta}{\partial x} - C_D U^2 \tag{5}$$

$$\frac{\partial Ch}{\partial t} + \frac{\partial CUh}{\partial x} = -v_m r_o C \tag{6}$$

$$(1-\lambda)\frac{\partial \eta}{\partial t} = I_f v_m r_o C \tag{7}$$

where h is the turbidity current depth, U is the depth-averaged underflow velocity, C is the depth-averaged volumetric concentration of mud, η is the elevation of the bottomset deposit (reservoir bed), t is time, x is the downslope distance along the bottom of the reservoir, e_w is the water entrainment coefficient, v_m is the mud fall velocity, δ is a parameter set equal to 1 in the muddy pond and set equal to 0 elsewhere, C_D is the bottomset friction coefficient, R is the submerged specific gravity of mud, and r_o is a constant that relates near-bed volumetric sediment concentration to the layer-averaged concentration.

In numerically solving Eqs. 1–3 and Eqs. 4–7 subject to the corresponding initial and boundary conditions, the advanced moving boundary framework of Kostic and Parker (2003a) with grid and time stretching was adopted. The fluvial and turbidity current sub-model were linked via two internal boundary conditions: the continuity condition at the foreset-bottomset break, and the shock condition that progrades the delta face (see Eqs. 2 and 4 in Kostic and Parker, 2003a).

In the turbidity current submodel, no bottom outlets were added at this point and the turbidity current can only vent when the water interface ξ_p in ponded zone rises above the crest elevation η_e of an uncontrolled or open-gate controlled spillway. In that case, the outflow boundary condition can be determined by combining the Bernoulli equation for the muddy underflow

$$RgC\big|_{x=L}\xi_p = \frac{1}{2}U_e^2 + RgC\big|_{x=L}(\eta_e + h_e) \tag{8}$$

with a densimetric Froude number of unity near the crest

$$U_e^2 = RgC\big|_{x=L}h_e \tag{9}$$

where U_e denotes the overflow velocity, h_e is the overflow depth and $C\big|_{x=L}$ is the volumetric concentration of mud in the pond just upstream of the dam. If, however, the water interface ξ_p in the muddy pond is below the overflow point η_e, the boundary condition is simply

$$U\big|_{x=L} = 0 \tag{10}$$

where $U\big|_{x=L}$ is the depth-averaged velocity in the pond just upstream of the dam.

The main advantage of the new formulation is a smooth and continuous progression from a no-overflow (stagnation) condition to an overflow condition at the dam. This is a serious computational challenge, which can only be overcome by a high-resolution, non-oscillatory numerical scheme, such as e.g. the Kurganov-Tadmore algorithm used here. In comparison, the model of Toniolo et al. (2007) with the McCormack scheme cannot continuously progress from a stagnation to an overflow condition. Rather, an appropriate boundary condition at the dam has to be selected by means of a trial and error procedure.

3   MODELING RESULTS

Two sets of flood events are evaluated herein. The first set of experiments (Figs. 2–5) aims at demonstrating the hydrodynamic response of the reservoir enclosed by a dam to a relatively short single-flood event. The second set of experiments (Figs. 6–7) was designed to quantify the long-term morphodynamic response of a reservoir to a series of intermittent flooding events.

The fluvial delta was assumed to have the following initial geometry: slope of the topset deposit $S_{fo} = 0.00073$, slope of the delta face $S_a = 0.2$, position of the foreset top $f_o = 1000$ m, elevation of the foreset top $\eta_{To} = 100$ m, elevation of the foreset toe $\eta_{Bo} = 0$ m. The remaining input parameters for the model were: water discharge per unit width $q_w = 6$ m²/s, volume sand feed rate per unit width $q_s = 0.0025$ m²/s, Chezy coefficient of topset bed $C_z = 12$, median diameter of sand $D_s = 500$ µm, sand porosity $\lambda_s = 0.4$, initial slope of the bottomset deposit $S_o = 0.01$, volume mud feed rate per unit width $q_m = 0.0225$ m²/s, bottomset friction factor $c_D = 0.001$, mud porosity $\lambda_m = 0.6$, submerged specific gravity of sand and mud $R = R_s = 1.65$, water surface elevation in the reservoir $Z_l = 103$ m, elevation of the spillway crest $\eta_{crest} = -400$ m, and position of the dam $L = 80$ km. Feed rates of both sand and mud were kept constant over time for simplicity.

The first set of experiments (Figs. 2–5) was performed with very fine mud (median diameter $D_m = 5$ µm) to facilitate the hydrodynamic response of a reservoir that is minimally influenced by the flow-reservoir bottom interaction. The experiment of Figure 2a illustrate well

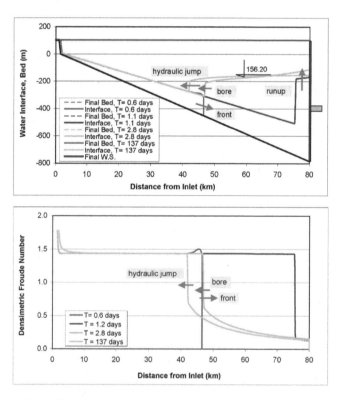

Figure 2.   Stages of turbidity current propagation and interaction with the dam: (a) turbidity current interface profile; (b) densimetric Froude number profile.

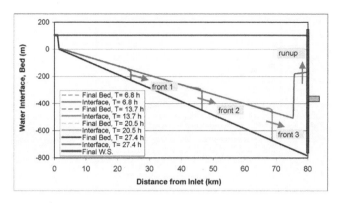

Figure 3.   Hydrodynamic response of a reservoir to single-flood events of varied duration.

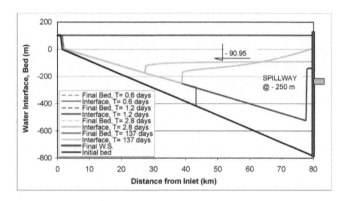

Figure 4.   Hydrodynamic effect of spillway crest elevation.

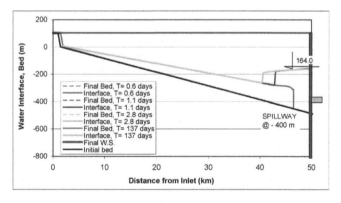

Figure 5.   Hydrodynamic effect of the dam location.

that the numerical model captures for the first time all stages of turbidity current propagation and interaction with a dam, including: a) the position of the current head progressing toward the dam; b) the runup against the face of the dam when the current collides with the barrier (stagnation point); c) the formation of an upstream-migrating bore as the current reflects off the dam; and d) the establishment of an internal hydraulic jump as the bore eventually stabilizes upstream. In this particular case, the jump establishes after 137 days of continuous flooding and the interface of the muddy pond stabilizes at 156.2 m. Figure 2b illustrates the

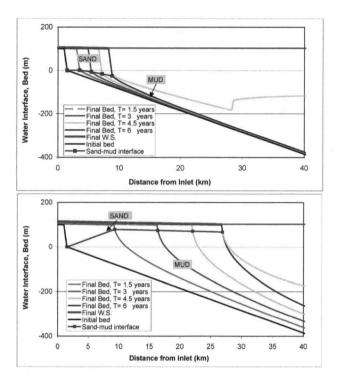

Figure 6.   Morphodynamic response of a reservoir to six years of continuous flooding: (a) $D_s = 5\,\mu m$; (b) $D_s = 20\,\mu m$.

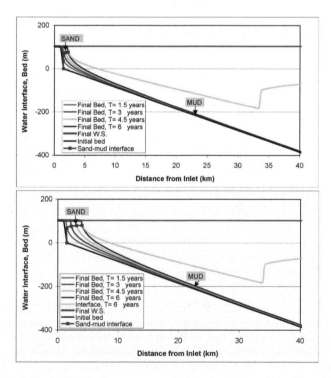

Figure 7.   Effect of flood intermittency on reservoir deposits and sand-mud interface: (a) $I_f = 0.05$; (b) $I_f = 0.1$.

variation of the densimetric Froude number along the reservoir during all stages of under-flow propagation and interaction with a dam. Clearly, the turbidity current is: a) Froude—supercritical as it progresses toward the dam; b) subcritical as it hits the dam and runs up against the dam face; and c) supercritical in the upstream part of the reservoir, subcritical in the ponded zone behind the dam (with very low densimetric Froude numbers), and transi-tional where a bore or a sharp hydraulic jump occurs.

The experiment of Figure 3 looks more closely into how the turbidity current of Figure 2 advances in time toward the dam. The results suggest that the current front does not reach the dam if flooding lasts less than approximately 20 h. As the turbidity current hits the dam, its head runs up against the barrier and the current starts venting over the spillway crest. Here the underflow reaches the dam after about 25 h of continuous flooding. The effect of the spillway elevation on the hydrodynamic response is analyzed by comparing the results of Figure 2 to those of Figure 4, which pertain to the same model parameters as the experiment of Figure 2, except for the value of the spillway elevation that was placed at −250 m in this case. Apparently, the supercritical current traveling toward a dam (Fig. 4) covers a shorter distance than in the previous case (Fig. 2) due to more significant accumulation of water mass behind the barrier, which limits the current propagation. However, a spillway placed higher is in a "no-overflow" (stagnation) mode longer, thus pushing the bore and internal hydraulic jump further upstream. The experiment of Figure 5 demonstrates the effect of the dam location. The model input parameters were identical to those of Figure 2 except for the distance from the inflow point to the dam, which was 50 km here. The hydrodynamic response of a shorter (and therefore shallower) reservoir is similar to that of a reservoir with a lower spillway crest, that is turbidity currents propagate further toward the dam, while the internal hydraulic jump does not get pushed so far upstream (Fig. 5).

The experiments of Figures 2–5 provide an insight into how high-resolution, non-oscillatory numerical modeling can be used to examine and quantify the hydrodynamic impact of a singe flood event on the reservoir performance, such as extreme floods associated with e.g. climate change and urbanization-induced land use/land cover changes. In addition to single-flood events, it is of value to analyze the morphodynamic response of a reservoir to a series of flood events and evaluate the long-term repercussions of reservoir sedimentation. This type of analysis can be crucial in selecting the appropriate routing or flushing operational strategies.

Here, the numerical experiments of Figures 6–7 illustrate the impact of multiple floods on the reservoir deposits and the interface of the muddy pond. Only first 40 km of the reservoir are shown in order to zoom in on the loss of reservoir storage in the vicinity of the river mouth. The overflow point in all experiment was kept at −250 m. The results of Figure 6a were obtained by running the experiment of Figure 4 for six years of continuous flooding (i.e. the flood intermittency I_f is assumed to be equal to 1). A noticeable fluvial delta emerges and starts prograding over and interact with a relatively thin bottomset deposit, forming a downward-facing sand-mud interface. In the experiment of Fig. 6b, the median size of mud of 5 μm was replaced with 20 μm in order to raise the interface in the muddy pond and increase the overspill by way of building the bottomset deposit more swiftly. An upward-fac-ing sand-mud interface indicates an increased rate of delta progradation due to a more rapid thickening of bottomset deposit and shortening of the drop of the sandy foreset deposit. The turbidity current interface in the muddy pond (not shown in Figure 6b) stabilizes at −90 m. Lastly, the experiment of Figure 6b was used as a basis for the analysis of Figure 7a, b on the effect of flood intermittency on resulting deposits and the position of the underflow interface relative to the spillway crest. The flood intermittency was set equal to 0.05 in the experiment of Figure 8a, and to 0.1 in the experiment of Figure 8b. This corresponds to about 18 days per year of cumulative flood time over a 6-year period in the former case, and to one month of flooding per year in the latter case. The inclusion of flooding intermittency results in more realistic deposits, that is the fluvial delta prograded at a lower rate and bottomset deposit builds up slower. The reservoir was venting over the spillway since the final interface in the muddy pond was found to be at about −51 m and −60 m respectively. i.e. over the crest kept at −250 m.

4 CONCLUSIONS

Numerical simulations with a high-resolution, non-oscillatory mass-conservative numerical model of reservoir sedimentation illustrate that this formulation can be successfully used to examine and quantify the hydrodynamic and morphodynamic impact of flooding on the reservoir performance. The model is, for the first time, capable of capturing all stages of turbidity current propagation and interaction with a dam, including: a) the position of the turbidity current front advancing toward the dam; b) the runup against the face of the dam as the current collides with the dam; c) the formation of an upstream-migrating bore as the current reflects off the dam; and d) the establishment of an internal hydraulic jump as the bore stabilizes upstream. This feature becomes crucial in the coming years when the climate-fueled extreme floods will continue to act to additionally reduce the reliability of reservoir flood control and other beneficial uses. The formulation will be amended in the next phase, so as to include bottom vents and allow for a variable water surface in the reservoir. With this modification, the model is expected to become a reliable predictive tool that can be used to evaluate and select the effective sediment routing, flushing and other operational procedures that could ensure sustained long-term use of reservoirs.

REFERENCES

Kostic, S. & Parker, G. 2003a. Progradational sand-mud deltas in lakes and reservoirs. Part 1. Theory and numerical modeling. *Journal of Hydraulic Research* 41 (2), 127–140.

Kostic, S. & Parker, G. 2003b. Progradational sand-mud deltas in lakes and reservoirs. Part 2. Experiment and numerical simulation. *Journal of Hydraulic Research* 41 (2), 141–152.

Kurganov, A. & Tadmor, E., 2000. New high-resolution central schemes for nonlinear conservation laws and convection-diffusion equations. *Journal of Computational Physics* 160, 241–282.

Lehner, B. et al. (2011). Global Reservoir and Dam (GRanD) database, Technical documentation, Version 1.1. http://www.gwsp.org/fileadmin/downloads/GRanD_Technical_Documentation_v1_1. pdf.

Leonard, B.P. 1991. The ULTIMATE conservative difference scheme applied to unsteady one-dimensional advection. *Computational Methods in Applied Mathematics and Engineering* 88, 17–77.

Morris, G. & Fan, J. 2010. Reservoir sedimentation handbook, Design and management of dams, reservoirs and watersheds for sustainable use. McGraw-Hill, New York.

Toniolo, H., Parker, G. & Voller, V. 2007. Role of ponded turbidity currents in reservoir trap efficiency. *Journal of Hydraulic Engineering* 133 (6), 579–595.

United Nations Environmental Programme. 2013. UNEP year book. Emerging issues in our global environment. UNEP Division of Early Warning and Assessment.

Reservoir Sedimentation – Schleiss et al. (Eds)
© 2014 Taylor & Francis Group, London, ISBN 978-1-138-02675-9

Innovative in-situ measurements, analysis and modeling of sediment dynamics in Chambon reservoir, France

M. Jodeau
EDF R&D LNHE, Chatou, France

M. Cazilhac & A. Poirel
EDF DTG, Grenoble, France

P. Negrello & K. Pinte
EDF CIH, Le Bourget du Lac, France

J.-P. Bouchard
Retired, formerly at EDF R&D LNHE, Chatou, France

C. Bertier
GIP Loire Estuaire, Nantes, France

ABSTRACT: This paper focuses on siltation processes in an Alpine reservoir, the Chambon Reservoir on the Romanche River. Suspended sediment concentration monitoring upstream the dam leads to the identification of the main contributing hydrological events. Downstream monitoring demonstrates that specific operating conditions (reservoir level, discharge) allow sediment routing throughout the reservoir. In order to elaborate a clear comprehension of sediment processes, field surveys have also been performed in the reservoir. Bathymetry, Velocity field, sediment concentration were monitored. An innovative device has been built in order to identify sediment and flow dynamics inside the reservoir. Some preliminary numerical simulation of sediment dynamics in the reservoir using TELEMAC2D and SISYPHE show encouraging results. Actually modeling could be a useful tool to evaluate sediment management strategy. The main processes involved in suspended sediment transport were identified and their understanding will help to define strategies to reduce sedimentation in Chambon reservoir.

1 INTRODUCTION

As it has been observed in many countries [[9]], sedimentation in reservoirs is unavoidable and may have several consequences: (i) loss of capacity, (ii) siltation near bottom gates, (iii) large sediment releases during reservoir emptying ...

In order to define long-term management of reservoir sedimentation, deposition in existing reservoirs needs to be mitigated by using appropriate measures for sediment release. The management of sedimentation in large reservoirs is a major issue. Indeed, large amount of fine sediments and gravels could deposit. In the case of large dams, flushing operations (opening of dam gates) could only venture turbidity current or erode a limited part of the sediment bed near the gates. It could require research worksto define the appropriate way of dealing with sediments in large reservoirs. For example, [[2]] studied turbidity current in Luzzone lake comparing 3D numerical calculations with in situ measurements; using laboratory experiments and numerical simulation [[11]] suggests to use geo-textile or underwater obstacle to deal with turbidity current, some numerical calculation were performed using Grimsel reservoir geometry, or [[10]] analyze the flow patterns and suspended sediment movement in pumped-storage facilities.

Before defining sediment operation, the main processes involved in sediment transport should be identified owing to measurements (bathymetric surveys, concentration monitoring, velocity measurements ...). They may help to identify the locations of deposition and the propagating ways (turbidity currents or homogeneous suspension).

EDF manages more than 400 dams. In several cases, sedimentation must be dealt with to avoid loss of storage or siltation near the bottom gates.

In this paper, we focus on the Chambon Reservoir, located in the Alps Mountains. We try to analyze the dynamics of sediment in the reservoir in order to be able to implement relevant sedimentation operations. In order to understand the dynamics of sediment, we analyze how sediment propagate through this large reservoir, first measuring sediment output and input, then we analyze the internal dynamics using in situ monitoring. The last part of this paper shows some preliminary numerical simulation of sediment dynamics in the reservoir using TELEMAC2D and SISYPHE. Actually modeling could be a useful tool to evaluate the sustainability of sediment management strategies.

2 SEDIMENTATION IN CHAMBON RESERVOIR

2.1 Description of dam and reservoir

The Chambon dam is located on the Romanche River in the French Alps. The watershed area at the dam is 254 km^2 and the elevation of the area is around 990 m. The Romanche river and two small water derivations flow into the reservoir, The Ferrand and Mizoen derivations, Figure 1. The hydropower facility, St Guillerme II, has been in activity since 1935, the head is 293 m and the electric power 110 MW.

The volume of water in the reservoir is estimated to be 47.5 10^6 m^3 and the reservoir is 3.5 km long at the highest water level. The water elevation varies depending on seasons, the water level fluctuations could be up to 60 m. Since the beginning of its use, the reservoir has undergone a high rate of sedimentation, it is due to the watershed geology, made of different areas of crystalline rocks but also metamorphic schist. The fine sediment deposition rate in the reservoir is around 100 000 m^3/year. In 2005, in order to protect the bottom gate of the dam, a dredging of 25 000 m^3 of sediments was performed.

The sedimentation in the reservoir is studied to find the best sustainable way to manage sediments.

2.2 Bathymetric analysis: Impact of water level regulation

Several bathymetries have been performed since the construction of the dam (six bathymetries from 1993 to 2011), they show that (i) from 1993 to 2006, a siltation rate of 100 000 to 200 000 m^3/year is measured; (ii) from 2006 to 2009, erosion of sediment is observed, it is due to the dredging near the bottom gate and to the low water level of the reservoir that induced an erosion of the upstream part of the reservoir. The difference between the last bathymetries

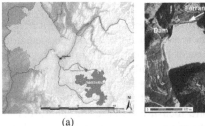

(a)   (b)

Figure 1.   (a) Location of hydropower facility Chambon-St Guillerme II. (b) Aerial picture of the reservoir, the water fall created by the Ferrand derivation (intake's outlet) is on the right bank.

(2009 and 2011, Fig. 2) highlights the significant role of specific water elevation due to particular reservoir operations:

- erosion is observed in the upstream part of the reservoir, in the areas higher than 1005 m. During this period (2009–2011) water level was always lower than 1005 m;
- deposition is measured between 985 and 995 m, where the reservoir is the largest. This could be explained by the fast deposition of coarse material;
- then erosion is shown between 985 and 980 m, 980 is the minimal turbinable water level, and 985 m is a water level that is not often reached;
- a deposited delta of sediment is observed below 980 m.

From the bathymetric data, we could conclude that the reservoir bed evolution is strongly impacted by the water level in the reservoir and its geometry: sediment are eroded in the upstream part of the reservoir where the water flows with high velocities and low water depths; sediment are deposited in the dam vicinity of the reservoir where the water is still and where the water depth could be high; the enlargement impacts the deposition/erosion processes.

2.3 Sediment characterization

Sediment were sampled from the bed in 2004, d_{50} is around 50 µm, and the concentration of the bed varies from 900 to 1200 g/l. The content of organic matter is low (around 2%). Due to their small grain size, these sediments are cohesive. Sediment fall velocity measurements have been performed in the laboratory on a representative sample of suspended sediments (d_{10}=3.7 µm, d_{50} = 10.9 µm, d_{90} = 37.9 µm). Settling velocity have been measured owing to a Andreasen pipette, a sediment weight device [[8]] and a SCAF device [[14]]. The data from three devices indicate the same trend: the settling velocity is the highest, 0.4 mm/s, for a concentration of 10 g/l, this value is much higher than the one that could be calculated owing to Stokes formula, 0.12 mm/s. For concentrations higher than 10 g/l a hindered regime is measured [[3]].

2.4 Analysis of sediment input and output

Continuous measurement of sediment concentration has been performed at several locations in order to characterize sediment input and output. Turbidimeters are located: (i) 4 km upstream the dam in the Romanche River; (ii) in the Ferrand derivation; (iii) downstream the reservoir in the Clapier Bassin. Turbidimeters (Hachlange SC100) are calibrated using samples, the calibration relation could be difficult to establish, therefore concentration measurements should be considered precociously.

The incoming sediment flux in the reservoir is calculated as the sum of the main Romanche input, both derivation inputs, and input from the reservoir subwatershed (between the upstream measurement site and the dam, the ratio of drainage area is 256/220).

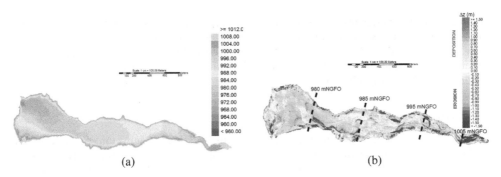

(a) (b)

Figure 2. (a) Bed elevation, data from 2011 bathymetric survey. (b) Measurement of bed evolution in the reservoir: difference between 2009 and 2011 bathymetries.

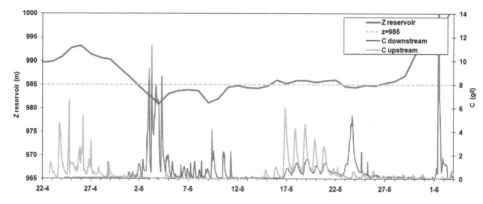

Figure 3.   Measurement of sediment concentration upstream and downstream the dam, example of 2004.

They indicate that the annual mean input in the reservoir is around 120 000 t, 78 000 t come from the upstream river, 8 100 t from the intermediate watershed and that the rest comes from the Ferrand derivation. From this quantity of sediment, 25% are transported downstream the dam and 75%, i.e. 90 000 t, deposit in the reservoir. These measurements also indicate the time variation of the sediment input: 50% of the sediment from the Ferrand derivation flow during 3% of the year and it corresponds to 4% of the global volume of water.

Furthermore, The input of sediment in the reservoir is controlled by daily snow-melt cycles, Figure 3, as it has been described by [[7]]. The output of sediment from the reservoir is strongly correlated to the water level in the reservoir, Figure 3 shows that: (i) above 985 m, sediment deposit in the upstream part of the reservoir; (ii) below 985 m, sediment are eroded and they could be transported to the water intake.

These bathymetric and concentration observations clearly show the part of the reservoir geometry in the sediment dynamics: the narrowing and the enlargement of the reservoir geometry impacts the deposition and erosion processes at the reservoir scale; water level in the reservoir is significant in the sediment dynamics.

Observation at low level of sediment transfer in the still area of water show that the transfer time is around 3 hours whereas the stay time is around 20 hours. This high difference could have two explanations: (i) sediment could be transported at high velocity near the bottom owing to turbidity currents; (ii) the Ferrand derivation and its water fall homogenize quickly the suspended sediments in the dam vicinity.

3   ANALYSIS OF INTERNAL SEDIMENT DYNAMICS

In order to have a better insight to the internal sediment dynamics, an innovative device has been designed. Its goal is to give continuous measurement of sediment concentration and flow velocities in the lake at a specific location. We wanted to have measurement in very different flow conditions. As the lake is very difficult to navigate in winter at low level, an autonomous device was chosen.

3.1   *Measurement of concentration and flow velocity in the reservoir*

The device was designed around a floating platform, two dead weights and a pulley system maintain the platform at the same location despite significant water elevation fluctuations. One turbidimeter (Hach Lange, solitax) was fixed near the bed (1 m) and another one near the water surface. A velocity measurement device (Nivus Doppler device) is fixed near the bed (1 m). The device has solar cells in order to be autonomous and data are uploaded owing to a GSM system.

Two platforms were put in the reservoir in order to compare sediment concentrations at two locations, Figure 4. The upstream one had some problems. When the water level had been very low, the platform touched down the sediment bed and whenthe water level rose again, around mid March, the platform stayed on the bottom under water, therefore it became out of service.

3.2   *Analysis of measurements from the platforms*

We presented the data for the first six months of 2013. Data should be analyzed regarding the incoming discharges and concentrations and the reservoir water level, Figure 5. In the beginning of 2013, the water level in the reservoir is very low due to work improvements on the dam. At several time the water level is lower than 985 m. Incoming discharges vary from 2 to 3 m^3/s from January to the beginning of April, then they become larger due to the snow melt, till 32 m^3/s. The incoming sediment concentrations are correlated to the incoming discharges, the higher measured value is 9 g/l the 11th May.

The measurements from the platforms present some lacks due to dysfunctions, Figure 6. Nevertheless they give information about sediment dynamics:

- the data from the top turbidimeter of the upstream platform could be analyzed till mid-March: value stay lower than 0.3 g/l but rises in concentration are observed when the water level is lower than 985 m;

(a)   (b)

Figure 4.   Autonomous platform to measure sediment flux in reservoir: (a) location of both platform in Chambon reservoir; (b) Photograph of the downstream platform.

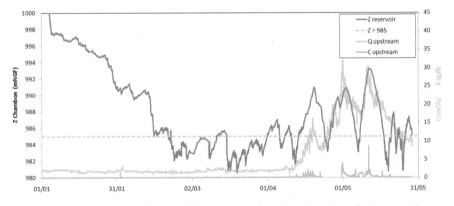

Figure 5.   2013 reservoir operating conditions: water elevation, incoming discharge and upstream sediment concentration.

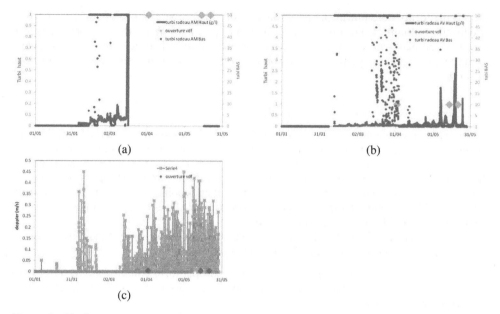

(a)

(b)

(c)

Figure 6. Platform measurements for the six first months of 2013: (a) concentration measurements at the upstream platform; (b) concentration measurements at the downstream platform; (c) velocity measurement near the bottom at downstream platform.

- at the downstream platform: measured velocities (near the bed) show a first increase to 40 cm/s when the water level became lower than 990 m. Concerning the concentrations: the data from the bottom turbidimeter could not be used from 2-02 to 10-03. From the beginning of April, data show strong correlation between the top and the bottom turbidimeters but with a difference of an order of magnitude (concentration around 3.5 g/l near the surface and around 15 g/l near the bottom). Two rises in concentration of May (6 and 9th) are not correlated with a increase of incoming discharge but with a lowering of the water level below 985 m.

4 NUMERICAL MODELING OF SEDIMENT DYNAMICS IN CHAMBON RESERVOIR

Numerical modeling could be a relevant tool in order to test sediment management strategies. In the specific case of large reservoir, 1D models have been used to predict the sediment concentration during lowering operations [[6]]. But due to the complex geometry of the large reservoirs, and stratification processes, 2D or 3D model could be required. 3D numerical modeling is now used to study reservoir sedimentation [[2, 13]]. We plan to build a 3D model of Chambon reservoir, first we analyze in this paper the results of some preliminary 2D calculations. Therefore, in this paper, we don't focus on 3D stratification in the dam vicinity of the reservoir.

4.1 Brief presentation of numerical tools and description of the model

Tools from the open source Telemac system (www.opentelemac.org) are used: TELEMAC 2D [[4]] for the hydrodynamic part of the calculation and SISYPHE [[12]] for the modeling of the sediment transport processes, that is to say in this case transport by suspension, erosion and deposition of cohesive materials. TELEMAC 2D solves the shallow water equations using either a finite element or finite volume schemes. Sisyphe solves a dispersion-advection equation with Krone and Partheniades formulas for respectively deposition and erosion flux (source terms).

The geometry of the model is based on the last bathymetry (2011, 1 point/m), the area of the model is the area under water for a water level of 1018 m (maximal operating level during the last years). We built a 50 m upstream prolongation of the model in order to avoid problems with the upstream boundary condition.

The mesh is made of triangular elements with a size of 5 m everywhere but in the talweg where there have a size of 2 m and near the bottom gate and the water intake. This mesh has 99 479 nodes and 197500 elements. HYPACK [[5]] and ape BLUEKENUE [[1]]softwares were used to build the mesh.

The upstream hydraulic condition is a varying discharge and the downstream condition could be an imposed water level or an output discharge. The concentration of sediment is chosen on the upstream boundary and it is a free condition on the downstream boundary. The water intake which is not at the boundary line, is represented by a sink.

No data is available to calibrate the friction coefficient, therefore the Strickler coefficient is chosen equal to 52 m$^{1/3}$s^{-1}. The turbulence chosen model is a constant viscosity (D = 10^{-3}m^{2}s^{-1}).

In these first calculations, a simple configuration is designed according to the measurements, the sediment bed is made of 2 m of uniform cohesive sediments, concentration of the bed is fixed to 900 g/l, fall velocity is 0.4 mm/s. Due to a lack of measurements, other parameters are chosen by analogy with other similar studies [[13]], that is to say: lateral and longitudinal diffusivity; critical shear stress for erosion τ_{CE} =1 Pa; Partheniades coefficient M =10^{-2} kg m^{-2}s^{-1}; and critical shear velocity for deposition v_{CD} = 0.01 m/s, equivalent to a critical shear stress of 0.1 Pa.

4.2 *Hydrodynamic results*

Calculations of hydrodynamics, using TELEMAC2D, have been performed first in order to identify the major areas in the reservoir and their possible role in the sediment transport dynamics. Figure 7 shows permanent results for an incoming discharge of 20 m³/s and a downstream water level of 980 m (minimal turbinable water level), water depth (a), scalar velocity (b), friction velocity (c) and Froude number are plotted.

Everywhere but in the still dam area, the water depths are small, often less than 1 m. In the upstream part, the water flows in a narrow channel (30–50 m wide) where flow velocities are

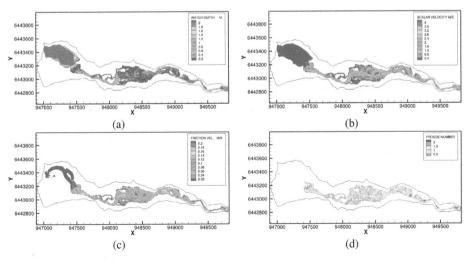

Figure 7.   Numerical results with TELEMAC2D for $Q_{upstream}$ = 20 m³/s et $Z_{downstream}$ = 980 m: (a) water depths; (b) depth average velocities; (c) shear velocities; (d) Froude number.

high, till 3 m/s. The downstream boundary of this narrow channel is located at the elevation of 999.35 m. According to Shields law about the shear stress for the beginning of motion, in these hydraulic conditions 3 cm gravels move. Downstream, in the enlargement, the talweg gets larger and the flow is braided and velocities decrease. The downstream boundary of this braided and large area is located at the elevation 985 m. Then, downstream the enlargement, there is another narrow channel where flow velocities are higher than 1 m/s. A typical pattern of sediment delta could be seen at the downstream boundary of this narrow channel. Near the dam, large water depths, low velocities are observed. In both narrow channels, the flow is super critical.

These calculations could explain the observations, reported paragraph 2.4, which highlight the key role of the enlargement: (i) above 985 m, sediment deposit in the upstream part of the reservoir (ii) around 985 m, sediment deposit in this area and (iii) below 985 m, sediment are eroded and they could be transported to the water intake.

4.3   Sediment transport numerical results

Calculations using SISYPHE are first performed to compare sediment settling for several water level in the reservoir (z = 1005,985 and 980 m). The incoming discharge is set to 20 m³/s which is typical of a snow melt discharge of the upstream Romanche River (Fig. 5), the upstream sediment concentration is set equal to 1 g/l. No bed evolution is taken into account, that is to say no erosion of the bed and deposition does not modify the bed elevation. We compare results for a computation that allows the permanent state to be reached, around 40 000 s.

Figure 8 (a,b,c) show the concentration in the reservoir in the case of 1005, 985 and 980 m water levels at the dam. Figure 8 (d) gives the concentration along the talweg in function of the distance to the dam. These figures show that

- whatever the downstream water level is, the sediment are convected in the narrow upstream channel with a concentration near the one imposed at the upstream boundary;
- for the lowest water level (985 and 980 m), we see some deposition in the secondary channels of the braided enlargement;
- as it could have been foreseen, deposition mainly occurs in the still downstream zone;
- the lower the downstream water level is, the higher the concentration at the dam is: the calculations give 0, 0.05 and 0.35 g/l at the dam for the respective water level of 1005, 985 and 980 m.

These values are coherent with the concentrations measured by the platform.

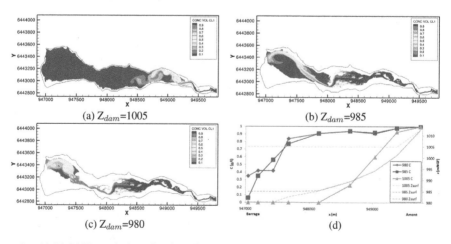

Figure 8.   (a)(b)(c) Numerical results of settling calculations: $Q_{upstream} = 20$ m³/s et $Z_{dam} = 1005$-985-980 m: sediment concentration. (d) sediment concentration (plain lines) and water elevation (dot lines), x is the distance from the dam.

5 CONCLUSION AND PERSPECTIVES

This paper focuses on the understanding of the sedimentation in the Chambon Reservoir. Analysis of input and output flux indicates that the deposition rate strongly depends on the water level in the reservoir. About the sediment dynamic in the reservoir, measurement data highlight the specific role of the geometry of the reservoir: the enlargement induced deposition and downstream an erosion zone could be observed. The mixing of water induced by the water fall of the Ferrand derivation is also a key process that forced sediment in suspension in the whole water depth in the dam vicinity. An innovative platform has been designed to allow a better insight in sediment concentration and flow velocity in the reservoir all over the 2013 year. More data will help to clearly identify the part of Ferrand water fall on the hydrodynamics in the lake.

Measurements data are compared with numerical calculations using TELEMAC2D and SISYPHE. Even if these calculation are very improvable they show a good agreement between numerical results and measurement. This 2D model could not be yet used to test sediment management strategies but it encourages further developments, for example 3D tests or more detailed description of the sediment bed. Real events of sediment transfer will be simulated and compared to in situ data. Then a 3D hydrodynamic and sediment model will be built using TELEMAC3D. A 2D model is satisfying in the upstream part of the reservoir where the velocities are high, the water depths low and the concentration homogeneous. In the downstream area, using a 3D model is the only way to reproduce stratification.

Furthermore, improvement of sedimentation management will be implemented using observation and modeling, it could be based on water level strategy.

REFERENCES

Blue kenue: Software tool for hydraulic modellers, http://www.nrc-cnrc.gc.ca, 2013.
G. De Cesare. *Alluvionnement des retenues par courants de turbidit*. PhD thesis, EPFL, 1998.
N. Gratiot, H. Michallet, and M. Mory. On the determination of the settling flux of cohesive sediments in a turbulent fluid. *Journal of Geophysical Research*, 110, 2005.
J.M. Hervouet. *Hydrodynamics of free surface flows, modelling with the finite element method*. Wiley, 2007.
Hypack. Hydrographic survey software. Technical report, Hypack Inc., http://www.hypack.com, 2013.
M. Jodeau and S. Menu. Sediment transport modeling of a reservoir drawdown, example of Tolla reservoir. In *River Flow Conference*, 2012.
V. Mano, J. Nemery, and A. Poirel. Assessment of suspended sediment transport in four alpine watersheds (France): influence of the climatic regime. *Hydrological Processes*, 23, 2009.
A. Mantovanelli and P.V. Ridd. Devices to measure settling velocities of cohesive sediment aggregates: A review of the in situ technology. *Journal of Sea Research*, 56 (3):199–226, 2006.
G.L. Morris and J. Fan. *Reservoir Sedimentation Handbook*. McGraw-Hill, 1997.
M. Muller. *Influence of in- and outflow sequences on flow patterns and suspended sediment behavior in reservoirs*. PhD thesis, EPFL LCH, 2012.
C.D. Oehy, G. De Cesare, and A.J. Schleiss. Effect of inclined jet screen on turbidity current. *Journal of Hydraulic Research*, 48:81–90, 2010.
P. Tassi, C Villaret, and D. Pham Van Bang. Sisyphe, release 6.2, user manual H-P73-2010-01219, available on www.open-telemac.org. Technical report, EDF, 2013.
E. Valette, P. Tassi, M. Jodeau, and C. Villaret. St Egrve Reservoir—Multi-dimensional modelling of flushing and evolution of the channel bed. In *River Flow*, 2014.
V. Wendling, Gratiot, C.N. Legout, I.G. Droppo, A.J. Manning, G. Antoine, H. Michallet, and M. Jodeau. A rapid method for settling velocity and flocculation measurement within high suspended sediment concentration rivers. In *Intercoh*, 2013.

Reservoir Sedimentation – Schleiss et al. (Eds)
© *2014 Taylor & Francis Group, London, ISBN 978-1-138-02675-9*

Measurements of spatial distribution of suspended sediment concentrations in a hydropower reservoir

S. Haun
Department of Hydraulic and Environmental Engineering, The Norwegian University of Science and Technology, Trondheim, Norway

L. Lizano
Design Department, Instituto Costarricense de Electricidad, San José, Costa Rica

ABSTRACT: Sediments carried by rivers begin to settle when the water enters a reservoir, resulting in sediment depositions. The aim of this study was to increase the knowledge of suspended sediment transport in a comparatively small reservoir located in the North-West of Costa Rica. Vertical distributions of suspended sediment concentrations and particle sizes were measured along the reservoir by using a LISST instrument in order to get spatial variations. The results showed that sediment concentrations and mean grain sizes decrease continuously along the reservoir, especially in the upstream area of the reservoir. A significant increase in the concentrations with depth could only be seen in the most upstream transect, all other transects showed an almost uniform distribution of the concentrations over the depth.

1 INTRODUCTION

A natural trapping of transported sediments happens when a river enters a lake or reservoir and the flow velocities and turbulences decrease. Worldwide about 1% of the storage volume is lost due to sediment depositions annually (Mahmood 1987). However, in areas with a distinctive rainy and dry season, like in Central America, the depositions can reach values up to 5% (Haun & Olsen 2012, Haun et al. 2013).

These depositions may result in a significant decrease of the reservoir lifetime, a shortage of electrical output, the loss of water for drainage purposes and often also in a loss of flood control benefits (Scheuerlein 1990, Morris & Fan 1998). Management tasks are therefore necessary to avoid negative effects on the operation of the reservoir and/or to avoid that the reservoir, in the worst case, is even lost after only a few years of operation. Yet, many reservoirs have been designed and built without appropriate knowledge about sediment inflow and about the expected sediment problems. As a consequence, many of them have faced sedimentation problems after starting the operation, which may limit the possible actions for sediment management (e.g. without a sufficient size of the bottom outlets, a reservoir flushing is not feasible anymore). However, if the reservoir is already constructed, it is possible to conduct measurements e.g. of the Suspended Sediment Concentrations (SSC) or the Particle Size Distribution (PSD) within it. Knowledge regarding the sediment transport processes in the reservoir makes it possible to accurately plan further sediment management tasks. Next to knowledge regarding the SSC, information about the PSD is also valuable, e.g. close to the intakes to get an idea of the characteristics of the sediments entering the system.

A hydropower reservoir in Costa Rica was chosen for this study. In this country, the high quantity of inflowing sediments is a result of the combination of the local geology, the topography of the catchment area and the high precipitation. Suspended sediment concentration and particle size distribution measurements in the reservoir were conducted by using a Laser In-Situ Scattering and Transmissometry instrument (LISST). The instrument is based on a laser-diffraction method and measures simultaneously sediment concentrations as well as the particle size distribution with a measurement frequency of 0.5 Hertz (Sequoia Scientific Inc. 2011). An advantage of the instrument is that no further calibration is necessary e.g. compared to a turbidity sensor (Agrawal & Pottsmith 2000). This device can measure 32 particle sizes in the range between 1.9 and 381 μm, which is another advantage for using it, if a high amount of fine sediment enters the reservoir. The LISST measurements were carried out in transects along the reservoir and over the depth of different verticals. The results were evaluated to give insight into the spatial distribution of the SSC and the PSD. Additionally, measurements with an Acoustic Doppler Current Profiler (ADCP) were conducted to help to understand the flow characteristics in the reservoir. The results of the LISST measurements are presented and discussed in this study.

2 BACKGROUND/FIELD SITE

2.1 Aim of the study

The aim of this study was to get better knowledge regarding the suspended sediment transport in the Peñas Blancas reservoir in Costa Rica, especially the 3D distribution of the SSC as well as the distribution of mean Particle Sizes (PS) in this reservoir. This knowledge may be used in additional studies to develop and improve new methods, like the use of ADCP backscatter data for a standardized evaluation of the suspended sediment load and as input for numerical models.

During the field investigations, the conditions in the reservoir changed frequently and so small concentrations, which are close to the limitation of the used instrument (LISST-SL) also occurred. Hence, it was another challenge to see if the instrument, which was already used in the Angostura reservoir in Costa Rica during high sediment laden inflow (Haun et al. 2013), is able to produce reliable results during low sediment transport conditions.

2.2 Field site

The field data were collected in the reservoir of Peñas Blancas power plant, which is located in the North-Western part of Costa Rica. The catchment, with 156 km², is covered mainly by forest and is subject to restrictions regarding the land use (ICE 1997). However, high sediment loads have been transported to the reservoir and have decreased its initial 3.32 million m³ volume over the time, mainly due to high precipitations—the average annual precipitation is 6066 mm (ICE 1997)—and through many landslides inside the catchment. Very soon after the project was built, it became evident that feasibility studies underestimated the sediment volume that would affect the power plant. Flushing with the bottom outlet, which was foreseen to preserve the reservoir volume, has been limited, due to environmental concerns. Actually, only partial flushings, with pressurized flow through the bottom outlet, are carried out.

The reservoir area at the highest regulated level is 350000 m² (ICE 1997). Water flows from the upstream part of the reservoir to the intake, located at the left side of the dam and with a maximum capacity of 33 m³/s. During the field measurements, high variations of the flow path were observed, depending on the inflow and the reservoir conditions.

The average discharge in the Peñas Blancas river is 23 m³/s, at the dam site (ICE 1997). The measurements took place during November and December 2013, when the Peñas Blancas river shows higher average discharges and empirically the largest part of the sediments enters the reservoir. Actually, discharges reached at some days, values higher than 100 m³/s. For the whole evaluation of the reservoir about 372 measurements were conducted within two weeks. 16 of the measurements conducted during the 19th December 2013 are presented in this study.

3 THE LISST PRINCIPLE, LISST-SL

For measuring the suspended sediment concentrations and the grain size distributions in the reservoir, a Laser In-Situ Scattering and Transmissometry instrument was used (Sequoia Scientific Inc. 2011). The LISST-SL device, used for this study, is Stream Lined (SL), so it adapts to the flow conditions also when the flow velocities are small. The instrument is based on the principle of laser diffraction and measures continuously suspended sediment concentrations as well as the particle size distribution and shows instantly results to the user. The LISST device consists in principle of a collimated Laser beam (*L*), a Receive lens (*R*), a multiring Detector (*D*) and a transmission detector (*P*; Fig. 1; Agrawal & Pottsmith 2000). The collimated laser beam (670 nm wavelength) is sent through Water (*W*), with a predefined path Length (*l*). Small particles which pass through the sample and cross the collimated laser beam scatter light at a particular angle to the scattered-light receiving lens which is then printed on a silicon multi-ring photo-detector (Agrawal & Pottsmith 2000, Agrawal & Traykovski 2001). The recorded scattering intensity is used during post-processing, where due to the use of the mathematical inversion, the power distribution may be converted to a size distribution. The volume concentration is subsequently calculated by multiplying the area in any size by the grain size. An advantage of the instrument is that the measurements are independent of the refractive index of the particles (Agrawal & Pottsmith 2000). Shape effects of natural sediment grains are included in processing of the results in the software, but have to be handled with care due to a dependency on the particle type (Agrawal et al. 2008, Felix et al. 2013).

In total, the measured PSD is subdivided into 32 grain sizes in the range between 1.9 and 381 μm. The pre-defined measurement frequency of the instrument is 0.5 Hertz. The maximum and minimum SSC which can be measured by the instrument depends in principle on the particle size (*d*) and on the path Length (*l*) of the collimated laser beam and is roughly *d/l* (Agrawal et al. 2011). While in the user's manual the maximum concentration that can be measured is given to be 2000 mg/l, the minimum concentration, which is very important for this study, is in the range of 10 mg/l (Sequoia Scientific Inc. 2011). The instrument has in addition an integrated depth and temperature sensor.

4 FIELD MEASUREMENTS

The measurements were applied in vertical profiles in 5 transects along the reservoir to get an overview of the SSC and the PSD within the reservoir (Fig. 2). The measurement interval in the vertical direction was chosen to be 2 m. The first measurement was conducted in a depth of 2 m due to the fact that floating material and small organic material was observed at the

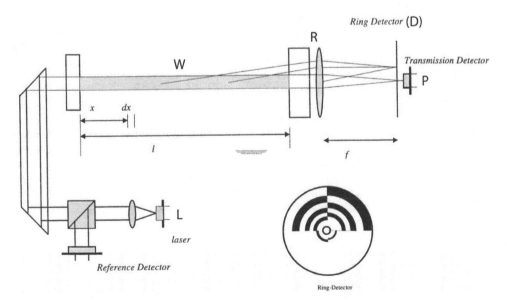

Figure 1.   Principle of the LISST measurement device (Agrawal and Pottsmith 2000, modified).

65

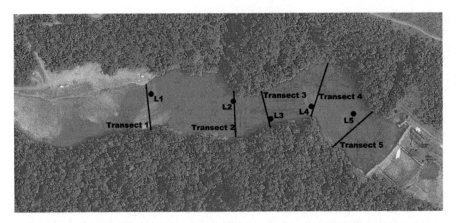

Figure 2. Peñas Blancas reservoir, with the 5 chosen transects and the vertical points were the LISST-SL measurements were conducted (L1–L5).

surface, and so it was assumed that measurements closer to the surface could be influenced by these particles.

The measurements were performed and processed as a time series and an average value for the SSC and the mean grain size was calculated for each point measurement. As example, transect 5 measured concentrations are shown in Figure 3. The results of the measurements are presented in Table 1, where the concentrations were subdivided accordingly to the sediment particle sizes, clay and fine silt (2 μm–6.3 μm), medium silt (6.3–20 μm), coarse silt (20 μm–63 μm) and fine sand (63 μm–200 μm), and shown as percentage of the total concentration. In addition a cumulative grain size distribution was processed for each point, again as average of the 30 second measurement. Figure 4 shows the cumulative grain size distribution, for the same vertical as shown in Figure 3. From Figure 4, it can also be seen that the cumulative grain size distribution does not start at zero. This is due to the fact, that the measured clay content in the water sample is added to the first measurable sediment size by the device.

In the field, a sensitivity analysis was performed to find an optimum duration of a measurement. The evaluation of the data showed that a measurement with 15 seconds at a frequency of 0.5 Hertz would be sufficient to give a reliable value. The deviation compared to a 60 second measurement is in the range of about 0.5 mg/l for a 15 seconds measurement and in the range of 0.35 mg/l for a 30 seconds measurement, which is comparable with the noise of the instrument. For this study finally a 30 seconds measurement interval was chosen.

In addition a 120 seconds noise test was performed, which answers the question if the device is adapted to the conditions at the field site (e.g. to the temperature) and if the collected background can be used for evaluating the measurements. The results showed that the noise may produce a change in the concentration of up to 0.32 mg/l, which proofs the usability.

What can happen during measurements in a reservoir and also in a river, are natural fluctuations in the concentrations due to e.g. a change in the flow conditions. The advantage of the LISST-SL is that the user can see the data in real time, which means that natural fluctuations of the concentrations in the reservoir that may occur suddenly can be observed directly and the measurements can be adapted immediately, e.g. a longer measurement period per point can be chosen. However, during these measurement a sudden in—or decrease was not observed.

4.1 Transect 1

Measurements were performed in a depth of 2 m and 4 m in this transect. An increase in the SSC from 2 m (26.15 mg/l) to 4 m (33.61 mg/l) could be recognized. Also, the mean PS showed a higher value in 4 m depth (33.54 μm), compared to the measurement in 2 m depth (30.66 μm). A more detailed look into the different grain sizes showed that the main percentage of the inflowing sediments (about 75%) is medium and coarse silt, with a considerable higher

66

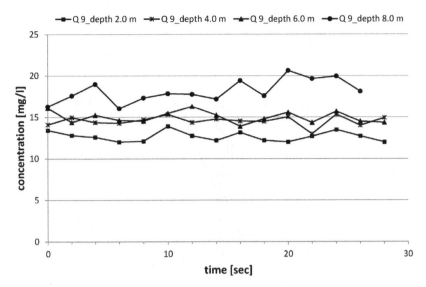

Figure 3.   Time series of the suspended sediment concentration measurements in the vertical in transect 5 in four different depths.

Table 1.   Overview of the measured suspended sediment concentrations, the calculated mean diameter and the percentages of the measured sediment fractions (clay, silt and sand).

Transect	Depth [m]	Concentration [mg/l]	Clay and fine silt [%]	Medium silt [%]	Coarse silt [%]	Fine sand [%]	Mean diameter [µm]
1	2.0	26.15	13.09	29.10	45.77	12.05	30.66
1	4.0	33.61	9.92	25.88	49.38	14.83	33.54
2	2.0	22.32	16.40	34.68	41.70	7.22	25.50
2	4.0	21.77	16.95	34.56	46.72	1.76	23.26
2	6.0	22.25	16.53	33.89	47.20	2.39	23.40
3	2.0	17.35	18.59	41.31	39.86	0.24	17.82
3	4.0	17.74	18.42	38.72	42.83	0.03	18.13
3	6.0	18.54	17.54	37.56	44.76	0.14	18.96
4	2.0	14.35	30.01	45.31	24.65	0.04	13.38
4	4.0	14.91	27.89	44.40	27.61	0.10	14.33
4	6.0	15.22	26.56	43.11	30.27	0.06	15.05
4	8.0	16.51	23.46	40.96	35.55	0.03	16.62
5	2.0	12.59	29.52	47.60	22.84	0.04	12.66
5	4.0	14.54	24.85	42.60	32.52	0.03	15.44
5	6.0	15.00	25.36	41.14	33.40	0.09	16.07
5	8.0	18.16	22.51	41.26	36.00	0.23	17.01

percentage of coarse silt. In addition about 12 and 14.8% of the inflowing sediment is fine sand, in a depth of 2 m and 4 m, respectively. The coarsest measured grain size in this vertical was 153 µm. From the measurements it can also be seen that the relative content of clay, fine and medium silt of the total concentration is higher in a depth of 2 m, and decreases in a depth of 4 m.

4.2   Transect 2

In this transect, it can be seen that there is almost no increase of the SSC over the depth (22.32 mg/l in 2 m depth and 22.25 mg/l in 6 m depth). If the results from the 120 seconds

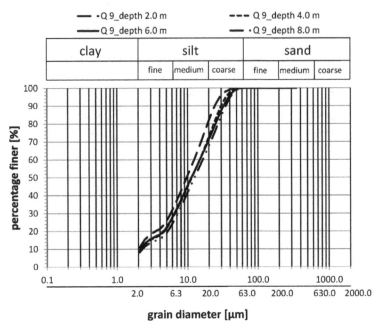

Figure 4.   Corresponding cumulative grain size distributions of the measurements presented in Figure 3.

noise test are taken into account (fluctuations in the range of 0.32 mg/l), then an almost uniform concentration of about 22 mg/l can be assumed over the depth. Medium and coarse silt are again the dominating fractions (with a percentage of 76.4–81.3%, together with fine sand, about 83%). In this section it can also be seen that the mean diameter in a depth of 2 m (25.5 μm) is slightly higher than the one measured in 4 m (23.26 μm) and 6 m (23.40 μm) depth. The measurements show that, in terms of absolute percentage by fraction, in a depth of 2 m the fine sand is about 5% higher than in the other two depths. Particularly, the measured grain size of 66.5 μm has a major influence on the high value of fine sand. A high amount of floating debris was observed in this area during the measurements. Hence, it may be assumed that small organic material, which was also observed close to the surface, was measured by the device in a depth of 2 m. ADCP measurements showed small flow velocities along the whole transect 2, so local turbulences which could keep fine sand particles close to the surface, can be excluded. A second measurement along transect 2, 50 meters from the first measured vertical, showed also a higher value for the fine sand in a depth of 2 m. Also in this area, the floating debris was observed at the surface. The coarsest measured grain size in this vertical was 130 μm. Compared to transect 1 (about 260 m upstream), a clear decrease of the SSC as well as of the mean PS can be seen (e.g. in a depth of 4 m, where the SSC dropped from 33.61 mg/l to 21.77 mg/l and the mean PS dropped from 33.54 μm to 23.26 μm).

4.3   Transect 3

Measurements in transect 3 show again almost similar concentrations over the depth (17.35 mg/l in 2 m and 18.54 mg/l in 6 m). Taking the noise into account, then the increase is in the range of only about 1 mg/l. Medium silt (41.3–37.6%) and coarse silt (39.9–44.8%) are in this transect, almost equally distributed over the depth and may be again characterized as the predominant sizes. The mean grain size increased only slightly from 17.82 μm in 2 m to 18.96 μm in 6 m depth. The coarsest measured grain size in this vertical was 78.9 μm. In addition, the percentage of fine sand does not play an important role anymore. Compared to transect 2 (about 130 m upstream), a decrease in the concentration could be recognized (e.g. in a depth of 4 m, where the SSC dropped from 21.77 mg/l to 17.74 mg/l). Additionally, a smaller mean PS was measured compared to the upstream located transect (dropped from 23.26 μm to 18.13 μm, again in a depth of 4 m).

4.4 Transect 4

The measurements showed a slight and continuous increase of the concentrations over the depth in transect 4. Again, if the noise is taken into account, then it can be seen that the concentration increase is in the range of about 1.5–2.8 mg/l within 6 meter depth (14.35 mg/l in 2 m depth and 16.51 mg/l in 8 m depth). The largest fraction of transported sediments is medium silt (40–45%). Furthermore in this transect for the first time a large decrease in the percentage of coarse silt (24.7–35.6%) could be seen. In contrast, the percentage of clay and fine silt (30–23.5%) increases significantly. If compared to transect 3 (about 130 m upstream), a decrease in the overall concentration (e.g. in a depth of 4 m, where the SSC dropped from 17.74 mg/l to 14.91 mg/l), yet not so big anymore, and a decrease in the mean PS (e.g. in a depth of 4 m, where the mean PS dropped from 18.13 μm to 14.33 μm), can be observed. The coarsest measured grain size in this vertical was 66.9 μm.

4.5 Transect 5

Measurement results in transect 5, the most downstream transect, are almost similar than those of transect 4, located about 140 m upstream. Concentration values (12.6–18.2 mg/l in a depth of 2 m and 8 m), mean grain sizes (12.7–17.0 μm in a depth of 2 m and 8 m) and the slight variation of these variables over the depth are similar in both transects. Comparing the percentages of the different fractions, a similar decrease of the clay and fine silt (29.5–22.5%) and medium silt fraction (47.6–41.3%) over the depth, and the increase in the percentage of the coarse silt (22.8–36%), is observed. Fine sand is in both transects too small, so it can be neglected. The coarsest measured grain size in this vertical was also 66.9 μm.

5 CONCLUSIONS/FUTURE WORK

The aim of this study was to increase the knowledge of suspended sediment transport in a small reservoir, regarding the spatial distributions and to get useful information about the settling of the particles in the reservoir. Measurements were performed with a LISST-SL device, which is based on the principle of laser diffraction. The measurements conducted during 19th December 2013 are presented in this study.

During this day, the measurements were performed in five transects along the reservoir, to see the changes in the concentration as well as the variation of the grain sizes. In each transect one vertical profile was chosen, with one measurement in each two meters of depth. The measurements were performed for 30 seconds in each point to get a time series of the concentrations and of the grain size distribution for the different points.

The main findings of this study can be summarized as follows. The suspended sediment concentrations in the reservoir and the mean grain sizes, which were calculated from the 32 measured particle sizes, decrease along the reservoir. Following the direction of flow, transect 1, the most upstream cross section, showed concentration values between 26.15 mg/l at the surface and 33.61 mg/l at the bed, and transect 5, close to the intake, showed lower values between 12.59 mg/l at the surface and 18.16 mg/l at the bed. Between transect 4 and 5 there was almost no more change in the concentrations, the grain sizes and the percentage of the different fractions (clay, fine silt, medium silt and coarse silt). The sediment size got finer from transect 1, where medium and coarse silt were the predominant fractions, to transect 5, where clay, fine and medium silt became more important. It could be seen that the percentage of fine silt and clay almost doubled along the reservoir. The fraction of fine sand decreased completely between transects 2 and transect 3. Transect 1, the most upstream cross section, showed values for the mean grain size between 30.66 μm at the surface and 33.54 μm at the bed, and transect 5, close to the intake, showed values between 12.66 μm at the surface and

17.01 μm at the bed. A strong increase in the concentration over the depth was only observed in the most upstream transect, all other transects had approximately uniform distributions of concentration and grain sizes over the depth.

The results presented in this study showed that the suspended sediment transport was very low during the measurements. Therefore, it can be assumed that they had very little influence on the sediment depositions in this reservoir. The measurements showed that it is feasible to use the LISST-SL device in reservoirs for evaluating the suspended sediment concentrations as well as the particle size distribution for small sediment concentrations. An additional conclusion from the field work is that the tests carried out for evaluating the noise ratio of the instrument for a certain field site and for certain conditions are extremely important. The noise ratio could be reduced to a value of only about 0.32 mg/l due to a careful collection of the background file.

A further step in this study would be to increase the resolution and to get a fully 3D pattern of the SSC and the PSD in the reservoir, by evaluating the backscatter data obtained by the ADCP. The results of this and other upcoming studies may be used as basis for further analyses, like for calibrating numerical models or for modifying the existing reservoir management tasks.

REFERENCES

Agrawal, Y.C. & Pottsmith, H.C. 2000. Instruments for Particle Size and Settling Velocity Observations in Sediment Transport, *Marine Geology 168*: 89–114.

Agrawal, Y.C. & Traykovski, P. 2001. Particles in the bottom boundary layer: Concentration and size dynamics through events, *Journal of Geophysical Research 106*(5): 9533–9542.

Agrawal, Y.C., Whitmire, A., Mikkelsen, O.A. & Pottsmith, H.C. 2008. Light scattering by random shaped particles and consequences on measuring suspended sediments by laser diffraction, *J. Geophys. Res., 113*.

Agrawal, Y.C., Mikkelsen, O.A. & Pottsmith, H.C. 2011. Sediment monitoring technology for turbine erosion and reservoir siltation applications, *Proc. HYDRO 2011 Conference, Prague, Czech Republic*.

Felix, D., Albayrak, I. & Boes, R.M. 2013. Laboratory investigation on measuring suspended sediment by portable laser diffractometer (LISST) focusing on particle shape, *Geo-Marine Letters 33*(6): 485–498.

Haun, S. & Olsen, N.R.B. 2012. Three-dimensional numerical modelling of reservoir flushing in a prototype scale, *International Journal of River Basin Management 10*(4): 341–349.

Haun, S., Kjærås, H., Løvfall, S. & Olsen, N.R.B. 2013. Three-dimensional measurements and numerical modelling of suspended sediments in a hydropower reservoir, *Journal of Hydrology, 479*: 180–188.

ICE, 1997. P.H Peñas Blancas. Estudio de factibilidad y diseño básico.

Mahmood, K. 1987. Reservoir sedimentation: impact, extent and mitigation, *World Bank Technical Paper 71*; ISSN 0253-7494, ISBN 0-8213-0952-8.

Morris, G.L. & Fan, J. 1998. *Reservoir Sedimentation Handbook*, McGraw-Hill Book Company, New York.

Scheuerlein, H. 1990. Removal of sediment deposits in reservoirs by means of flushing, *Proceedings International Conference on Water Resources in Mountainous Regions*, Symp. 3: Impact of Artificial Reservoirs on Hydrological Equilibrium, Lausanne, Switzerland.

Sequoia Scientific Inc., 2011. *LISST-SL User's Guide V 2.1*. Bellevue, USA.

Reservoir Sedimentation – Schleiss et al. (Eds)
© *2014 Taylor & Francis Group, London, ISBN 978-1-138-02675-9*

Case studies of reservoir sedimentation as a consequence of soil erosion

V. Hrissanthou
Department of Civil Engineering, Democritus University of Thrace, Xanthi, Greece

ABSTRACT: Two case studies for the computation of reservoir sedimentation are described in the present paper: Forggensee Reservoir (Bavaria, Germany) and Yermasoyia Reservoir (Cyprus). The physical processes that were quantified, are: runoff resulting from rainfall, soil erosion due to rainfall and runoff, inflow of eroded particles into streams, sediment transport in streams, sediment inflow into the reservoir, and sediment deposition in the reservoir. The above physical processes were included in a mathematical simulation model which computes the inflowing sediment quantity into the reservoir considered from the corresponding basin. The computed sediment inflow into Forggensee Reservoir was compared with available sediment yield (suspended load) measurements at the reservoir inlet for 12 years (1966–1977), while the computed sediment inflow into Yermasoyia Reservoir was compared with the measured mean annual rate of soil erosion in the corresponding basin. In both cases, the comparison results are satisfactory.

1 INTRODUCTION

Reservoir sedimentation, which decreases the useful life of the reservoir, is closely associated with soil and stream bed erosion in the corresponding basin. Sediment inflowing into the reservoir originates mainly from the products of soil and stream bed erosion. Therefore, the computation of reservoir sedimentation requires, in a previous step, the computation of soil erosion and stream sediment transport. The computation of soil erosion, in turn, requires the preliminary computation of runoff due to rainfall, because both rainfall and runoff cause soil erosion. According to the above mentioned process chain, the distinct physical processes, that should be quantified by means of corresponding distinct submodels, are: runoff resulting from rainfall, soil erosion due to rainfall and runoff, inflow of eroded particles into streams, sediment transport in streams, sediment inflow into the reservoir, and sediment deposition in the reservoir. The physical processes mentioned above could be included in a mathematical simulation model, which will compute the inflowing sediment quantity into a reservoir from the surrounding basins. For the verification of the mathematical model, usually, no systematic long-term sediment yield measurements of the rivers, streams or torrents, discharging their water into the reservoir, are available. Contrarily, rainfall and, in general, meteorological data, which serve as input data of the mathematical models, can be found. Additionally, model parameters, which serve also as input data, can be estimated by means of topographic and geologic maps.

In the following paragraph, the names of the reservoirs studied, the submodels used and the available output data are given:

Forggensee Reservoir (Bavaria, Germany)
– Rainfall-runoff submodel of Lutz (1984)
– Universal Soil Loss Equation (USLE)
– Stream sediment transport submodel of Yang and Stall (1976)

Available output data: sediment yield (suspended load) at the reservoir inlet for 12 years (1966–1977)

Yermasoyia Reservoir (Cyprus)
- Rainfall-runoff submodel (Giakoumakis et al. 1991)
- Soil erosion submodel of Schmidt (1992)
- Stream sediment transport submodel of Yang and Stall (1976)

Available output data: mean annual rate of soil erosion in the corresponding basin

2 FIRST MATHEMATICAL SIMULATION MODEL

The mathematical simulation model used in the case study of Forggensee Reservoir consists of four submodels: (a) a rainfall-runoff submodel; (b) a soil erosion submodel; (c) a sub-model for the estimation of sediment inflow into the main stream of a sub-basin; and (d) a submodel for the estimation of sediment yield at the outlet of a sub-basin.

2.1 Rainfall-runoff submodel of Lutz

The rainfall-runoff submodel of Lutz (1984) predicts rainfall excess for a given storm by using region-dependent and event-dependent parameters. Region-dependent parameters are the land use and the hydrologic soil group which reflects the infiltration rate.

The model of Lutz is expressed mathematically by the following equation:

$$h_o = (N - A_v)c + \frac{c}{k}[e^{-k(h_o - A_v)} - 1] \tag{1}$$

where h_0 = daily rainfall excess (mm); N = daily rainfall depth (mm); A_v = initial abstraction consisting mainly of interception, infiltration and surface storage and depending on the land use (mm); c = maximum end runoff coefficient expected for a rainfall depth of about 250 mm and depending on the land use and the hydrologic soil group; and k = proportionality factor (mm^{-1}) which is given by the following equation:

$$k = P_1 e^{-2.0/WZ} e^{-2.0/q_B} \tag{2}$$

where P_1 = region-dependent parameter; WZ = week number which designates the season; and q_B = baseflow rate which designates the antecedent moisture conditions (l s^{-1} km^{-2}).

2.2 Universal Soil Loss Equation (USLE)

The classical form of the USLE (Wischmeier & Smith 1978) is:

$$A = R \, K \, LS \, C \, P \tag{3}$$

where A = soil loss due to surface erosion (t ha^{-1}); R = rainfall erosivity factor (N hr^{-1}); K = soil erodibility factor (t ha^{-1} N hr^{-1}); LS = topographic factor; C = crop management factor; and P = erosion control practice factor.

The USLE is intended to estimate average soil loss over an extended period, e.g. mean annual soil loss (Foster 1982). However, only raindrop impact is taken into account in this equation to estimate soil loss. An improved erosivity factor was introduced by Foster et al. (1977) to take also into account the runoff shear stresses effect on soil detachment for single storms:

$$R = 0.5R_{st} + 0.5R_R = 0.5R_{st} + 0.5 \, a \, h_o \, q_p^{0.33} \tag{4}$$

where R = modified erosivity factor (N hr^{-1}); R_{st} = rainfall erosivity factor (N hr^{-1}); R_R = runoff erosivity factor (N hr^{-1}); h_o = runoff volume per unit area (mm); q_p = peak runoff rate per unit area (mm hr^{-1}); and a = a constant depending on the units ($a = 0.70$).

2.3 Estimate of sediment inflow into the main stream of a sub-basin

Sediment from the soil erosion transported to the main stream of a sub-basin is computed by the concept of overland flow sediment transport capacity. At this point, it must be noted that only the main stream of the sub-basin is considered because large amounts of unavailable data for the geometry and hydraulics of the entire stream system would otherwise be required.

The amount of sediment due to soil erosion transported to the main stream of a sub-basin, ES, is estimated by means of the following relationships:

$$ES = q_t, \quad \text{if} \quad q_{rf} > q_t \tag{5}$$

$$ES = q_{rf}, \quad \text{if} \quad q_{rf} \le q_t \tag{6}$$

where q_{rf} = available sediment in a sub-basin; and q_t = overland flow sediment transport capacity.

However, sediment from the preceding sub-basin, FLI, is also transported to the sub-basin under consideration. The total sediment transported to the main stream of the sub-basin, ESI, is therefore:

$$ESI = ES + FLI \tag{7}$$

The following relationships of Beasley et al. (1980) are used to compute the overland flow sediment transport capacity in a sub-basin:

$$q_t = 146 \, s \, q^{1/2} \quad \text{for } q \le 0.046 \text{ m}^3 \text{ min}^{-1} \text{ m}^{-1} \tag{8}$$
$$q_t = 14600 \, s \, q^2 \quad \text{for } q > 0.046 \text{ m}^3 \text{ min}^{-1} \text{ m}^{-1} \tag{9}$$

where q_t = overland flow sediment transport capacity (kg min^{-1} m^{-1}); s = mean slope gradient; and q = flow rate per unit width (m^3 min^{-1} m^{-1}).

Since the relationships of Beasley et al. (1980) are combined with USLE, the quantity q_{rf} is computed on the basis of the soil erosion amount A.

2.4 Estimate of sediment yield at the outlet of a sub-basin

The sediment yield at the outlet of a sub-basin, FLO, reflects the same basic controls as the sediment supply ES from soil erosion:

$$FLO = q_{ts}, \quad \text{if } ESI > q_{ts} \tag{10}$$
$$FLO = ESI, \quad \text{if } ESI \le q_{ts} \tag{11}$$

where q_{ts} = sediment transport capacity by streamflow.

The following relationships of Yang & Stall (1976) are used to compute sediment transport capacity by streamflow:

$$\log c_t = 5.435 - 0.286 \log \frac{w D_{50}}{v} - 0.457 \log \frac{u_*}{w}$$
$$+ \left(1.799 - 0.409 \log \frac{w D_{50}}{v} - 0.314 \log \frac{u_*}{w} \right) \log \left(\frac{us}{w} - \frac{u_{cr} s}{w} \right) \tag{12}$$

$$\frac{u_{cr}}{w} = \frac{2.5}{\log(u_* D_{50} / v) - 0.06} + 0.66, \quad \text{if } 1.2 < \frac{u_* D_{50}}{v} < 70 \tag{13}$$

$$\frac{u_{cr}}{w} = 2.05 \quad \text{if} \quad \frac{u_* D_{50}}{v} \ge 70 \tag{14}$$

where c_t = total sediment concentration (ppm); w = terminal fall velocity of suspended particles (m s^{-1}); D_{50} = median grain diameter of the bed material (m); v = kinematic viscosity of the water (m^2 s^{-1}); u_* = shear velocity (m s^{-1}); u = mean flow velocity (m s^{-1}); u_{cr} = critical mean flow velocity (m s^{-1}); and s = energy slope.

Equation 12 was determined from the concept of unit stream power (rate of potential energy expenditure per unit weight of water, us) and dimensional analysis. The variable u_{cr} in Equation 12 suggests that a critical situation is considered at the beginning of sediment particle motion, as in most sediment transport equations. But the relationship of Yang and Stall (1976) has the advantage, in contrast to other published equations, that it was verified in natural rivers.

3   SECOND MATHEMATICAL SIMULATION MODEL

The mathematical simulation model used in the case study of Yermasoyia Reservoir consists of four submodels equivalent to the ones applied to the basin of Forggensee Reservoir, namely: (a) a rainfall-runoff submodel; (b) a soil erosion submodel; (c) a submodel for the estimate of sediment inflow into the main stream of a sub-basin; and (d) a submodel for the estimate of sediment yield at the outlet of a sub-basin. The submodel (d) is exactly the same as in the first mathematical simulation model.

3.1   Rainfall-runoff submodel of water balance

The hydrologic submodel described in this section is a simplified water balance model (Giakoumakis et al. 1991), in which the variation of soil moisture due to rainfall, evapotranspiration, deep percolation and runoff is considered. The rainfall losses taken into account in the hydrologic submodel, are due to evapotranspiration and deep percolation. The basic balancing equation is:

$$S_n' = S_{n-1} + N_n - E_{pn} \tag{15}$$

where S_{n-1} = available soil moisture for the time step $n - 1$ (mm); N_n = rainfall depth for the time step n (mm); E_{pn} = potential evapotranspiration for the time step n (mm); and S_n' = auxiliary variable (mm).

The rainfall excess or direct runoff depth h_{on} (mm) and the deep percolation IN_n (mm) for the time step n can be evaluated as follows:

If $S_n' < 0$, then $S_n = 0$, $h_{on} = 0$ and $IN_n = 0$

If $0 \le S_n' \le S_{max}$, then $S_n = S_n'$, $h_{on} = 0$ and $IN_n = 0$

If $S_n' > S_{max}$, then $S_n = S_{max}$, $h_{on} = k(S_n' - S_{max})$ and $IN_n = k'(S_n' - S_{max})$,

where k and k' are proportionality coefficients ($k = 1 - k'$).

The maximum available soil moisture S_{max} (mm) is estimated by the following relationship of US Soil Conservation Service (SCS 1972):

$$S_{max} = 25.4 \left(\frac{1000}{CN} - 10 \right) \tag{16}$$

where CN is the curve number depending on the soil cover, the hydrologic soil group and the antecedent soil moisture conditions ($0 < CN < 100$).

3.2   Soil erosion submodel of Schmidt

The soil erosion submodel of Schmidt (1992) is based on the assumption that the impact of droplets on the soil surface and the surface runoff are proportional to the momentum flux contained in the droplets and the runoff, respectively.

The momentum flux exerted by the falling droplets, φ_r (kg m s^{-2}), is given by:

$$\varphi_r = Cr \, \rho \, A_E \, u_r \, \sin a \tag{17}$$

where C = soil cover factor; r = rainfall intensity (m s^{-1}); ρ = water density (kg m^{-3}); A_E = sub-basin area (m^2); u_r = mean fall velocity of the droplets (m s^{-1}); and a = mean slope angle of the soil surface (°).

The momentum flux exerted by the runoff, ϕ_f (kg m s^{-2}), is given by:

$$\phi_f = q\,\rho\,b\,u \tag{18}$$

where q = direct runoff rate per unit width (m^3 s^{-1} m^{-1}); b = width of the sub-basin area (m); and u = mean flow velocity (m s^{-1}).

3.3 Estimate of sediment inflow into the main stream of a sub-basin

Equations 5, 6 and 7 are also applied to the second mathematical simulation model. The available sediment discharge per unit width, q_{rf} (kg s^{-1} m^{-1}), due to rainfall and runoff, in a sub-basin is given by (Schmidt 1992):

$$q_{rf} = (1.7E - 1.7)10^{-4} \tag{19}$$

where

$$E = \frac{\varphi_r + \varphi_f}{\varphi_{cr}}(E > 1) \tag{20}$$

where ϕ_{cr} = critical momentum flux (kg m s^{-2}).

The critical momentum flux ϕ_{cr}, which designates the soil erodibility, can be calculated from:

$$\varphi_{cr} = q_{cr}\,\rho\,b\,u \tag{21}$$

where q_{cr} = direct runoff rate per unit width at initial erosion (m^3 s^{-1} m^{-1}).

The sediment transport capacity by overland flow, q_t (kg s^{-1} m^{-1}), is computed as follows (Schmidt 1992):

$$q_t = c_{max}\rho_s\,q \tag{22}$$

where c_{max} = concentration of suspended particles at transport capacity (m^3 m^{-3}); and ρ_s = sediment density (kg m^{-3}).

4 APPLICATION OF THE FIRST MODEL TO THE BASIN OF FORGGENSEE RESERVOIR

The first mathematical simulation model, that was described in Section 2, was applied to the 1500 km^2 basin of the Forggensee Reservoir (Bavaria, Germany), in order to compute the annual sediment yield at the basin outlet, namely at the reservoir inlet, for certain years. The storage capacity of the reservoir is 168×10^6 m^3. The largest part of the basin is in Austria and the main stream is the Lech River. The basin consists mainly of forest (36%), meadow (49%), and rock (11%) over 2000 m in altitude (Fig. 1, Hrissanthou 1990). Information about the soil texture class was available only for a small part of the basin where it consists of loamy sand, sandy loam, and clay loam. For more precise calculations, the basin was divided into 88 natural sub-basins, about 25 km^2 in area.

The following data were available: (a) daily rainfall amounts from five rainfall stations in the basin for 12 years (1966–1977); (b) suspended load at the outlet of the basin for these same 12 years, on a daily basis.

Sediment yield at the outlet of the basin was computed by the model on a daily basis because the rainfall amounts (input data) were available on a daily basis. Daily rainfall occurrences were treated as individual storm events. The daily values of sediment yield were added

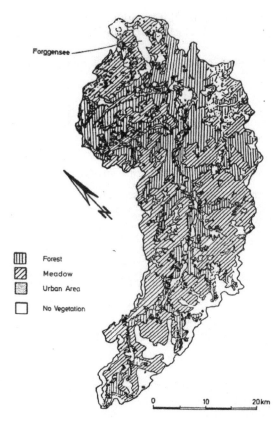

Forggensee

Forest
Meadow
Urban Area
No Vegetation

0   10   20km

Figure 1.   Soil cover map of the basin of Forggensee Reservoir (Hrissanthou 1990).

to produce the annual value of sediment yield at the outlet of the basin. Annual bed load was assumed to be 20% of the annual suspended load (Schröder & Theune 1984). The ratios of the computed annual values of sediment yield, associated with soil and stream bed erosion, to the measured values of sediment yield at the outlet of the whole basin are presented in Table 1.

4.1   Discussion—conclusions

The arithmetic results are satisfactory considering the large basin area and the fact that the computation was performed on a daily basis and that no runoff or sediment yield data were available for the sub-basins.

The rainfall-runoff submodel of Lutz (1984) is particularly suitable for the basin considered because it was developed for South German climatic conditions. The factors K, C and P of the empirical USLE were evaluated by means of the tables of Schwertmann (1981), which are valid especially for Bavaria. The factor R_{st} was estimated as a function of the daily rainfall amount on the basis of a regression analysis (Hrissanthou 1989).

Gully and bank erosion, as well as mass movement were ignored because no information was available to characterize them. Snowmelt runoff, glacial and snow erosion were not quantified because the research was focused on the classical soil erosion due to rainfall and runoff.

Moreover, USLE and any equation for classical erosion due to rainfall and runoff are not appropriate to be applied to rock areas without vegetation. As mentioned before, a part (11%) of the basin area consists of rock without vegetation, over 2000 m in altitude.

Table 1. Ratio of computed to measured values of sediment yield.

Year	Measured value (t)	Computed value/ Measured value
1966	585,600	1.46
1967	351,600	2.45
1968	374,400	1.78
1969	246,000	1.74
1970	1,165,200	0.88
1971	326,052	1.59
1972	79,046	5.30
1973	408,352	1.23
1974	324,037	2.48
1975	745,586	0.91
1976	315,772	1.36
1977	312,025	2.18

Finally, in this example, the comparison between computed and measured values of sediment yield at the outlet of the entire basin was made on an annual basis, although the calculations were performed on a daily basis. The following reasons are given for using an annual basis for the comparison: (a) the very long sediment travel times from the outlets of the most sub-basins to the outlet of the whole basin; (b) the fact that daily rainfall occurrences were treated as individual storm events; (c) the lack of runoff and sediment yield data in the sub-basins. These reasons render the precise computation of daily sediment yield at the outlet of the whole basin difficult. The addition of the daily values of sediment yield at the outlet of the basin causes a decrease in the differences between computed and measured values of sediment yield.

5 APPLICATION OF THE SECOND MODEL TO THE BASIN OF YERMASOYIA RESERVOIR

The second mathematical simulation model, that was described in Section 3, was applied to the basin of Yermasoyia Reservoir that is located northeast of the town of Limassol, Cyprus, in order to compute the annual soil erosion amount in the above basin, as well as the annual sediment yield at the basin outlet for certain years. The storage capacity of the reservoir is 13×10^6 m^3. The Yermasoyia River drains a basin that, upstream of the reservoir, amounts to 122.5 km^2. The length of the main stream of the basin is about 25 km, and the highest altitude of the basin is about 1400 m. The basin which consists of forest (57.7%), bush (33.7%), cultivated land (5.8%), urban area (1.8%) and an area with no significant vegetation (1%), was divided into four natural sub-basins for more precise calculations (Fig. 2, Hrissanthou 2006). The sub-basin areas vary between 14 and 44 km^2.

The soil types of the basin were divided into three categories: permeable (Calcaric Cambisols, Eutric Cambisols), semi-permeable (Eutric Regosols) and impermeable (Calcaric Lithosols, Eutric Lithosols) soils, because the above distinction is necessary for the estimate of the curve number in the hydrologic submodel.

Daily rainfall data for four years (1986–1989) from three rainfall stations were available. The mean annual rainfall at these stations amounts to 662 mm. Additionally, mean daily values of air temperature and relative air humidity and daily values of sunlight hours for the above four years were available from a meteorological station. Mean daily values of wind velocity only for one year (1988) were obtained from the same meteorological station. The air temperature, relative air humidity, sunlight hours and wind velocity data are necessary for the calculation of the potential evapotranspiration according to the radiation method improved by Doorenbos & Pruitt (1977).

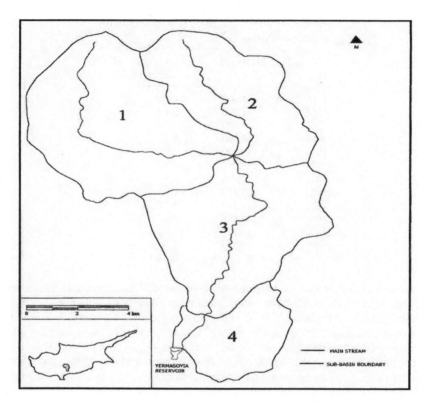

Figure 2.   Main streams of the four sub-basins of Yermasoyia Reservoir basin (Hrissanthou 2006).

Finally, the distribution of mean annual erosion rates over the island of Cyprus was obtained from the Water Development Department (Nicosia, Cyprus). According to this authority, the erosion rates have been deduced and assigned to the various geomorphologic areas of Cyprus on the basis of existing, randomly obtained, suspended sediment samples and mainly on the basis of estimates derived by surveying three dams.

The mathematical model was applied to each sub-basin separately and on a monthly time basis for a certain year. The monthly values of sediment yield at the basin outlet resulting from the model for a given year were added to produce the annual value of sediment yield YA due to soil and stream bed erosion. The annual soil erosion amount for the whole basin is symbolized with YD. The ratio of YA to YD is called the sediment delivery ratio DR. The computational results for YA, YD and DR for the years 1986–1989 are shown in Table 2.

The mean annual value of YD, 378,000 t, is transformed into the mean annual rate of soil erosion, 1.16 mm. The latter value is 1.7 times higher than the corresponding estimated value of 0.70 mm (Water Development Department, Nicosia, Cyprus). This estimated value is assigned to areas with igneous rocks, steep slopes and rainfall rates of the order of 600–800 mm/year, covered by forest, brush and with little cultivation. These climatic and physiographic conditions are fulfilled by the basin of Yermasoyia Reservoir.

According to the classical diagram of Brune (1953), the trap efficiency of Yermasoyia Reservoir is 100%. This means that all of the sediment yield at the basin outlet is deposited in the reservoir. Considering the storage capacity of the reservoir (13.6×106 m^3), its useful life thus amounts to 193 years.

5.1   *Discussion—conclusions*

The rainfall-runoff submodel of water balance (Giakoumakis et al. 1991) is suitable for the basin considered because it was developed for Mediterranean climatic conditions.

Table 2. Computational results for *YD, YA* and *DR.*

Year	*YD* (t)	*YA* (t)	*DR* (%)
1986	113,000	32,000	28
1987	673,000	224,000	33
1988	618,000	238,000	38
1989	108,000	30,000	28
Mean value	378,000	131,000	32

The soil erosion submodel, that is based on fundamental physical concepts, overestimates, finally, the mean annual rate of soil erosion for the whole basin. The following factors, amongst others, contribute to the deviation (overestimation or underestimation) between the computed and the measured values of soil erosion: (a) the equations for soil erosion described above were applied to relatively small sub-basins, whereas they were developed initially for small experimental fields; (b) snowmelt runoff, gully and bank erosion were neglected; (c) the erosion measurements are indirect, as explained previously.

The sediment inflow into the reservoir resulting from the stream sediment transport sub-model is also overestimated because the overestimated soil erosion quantity serves as input to the stream sediment transport submodel. Moreover, in the present case study, the bed of the main streams of the sub-basins consists of sand and gravel, and the slope of the main streams of two sub-basins exceeds the application limit of the Yang formula (1976), which is valid for sandy beds.

In the computations performed through the mathematical model, the sub-basin was the space unit and the month was the time unit. However, it has to be stressed that performing the calculations on an event basis is a reasonable way for the quantification of runoff, erosion and sediment transport processes, provided that pertinent detailed data are available.

In the case of large basins, for which mean annual values of soil erosion and sediment yield are required, the monthly time basis of the computations constitutes a temporally detailed approach. However, the performance of the calculations on a monthly time basis has as the consequence that some variables of the model equations, e.g. the rainfall intensity, characterizing single storm events lose their physical meaning.

Most parameters of the mathematical model used were estimated by means of tables, and topographic, vegetation and soil maps.

6 GENERAL CONCLUSION

From the preceding discussions on the computational results of the mathematical simulation models described above, it is concluded that these models are applicable to reservoir basins for which both hydrometeorological data, and topographic, vegetation and soil maps are available, in order to predict roughly reservoir sedimentation in terms of soil erosion.

However, it has to be emphasized that some of the model imperfections given above lead to an overestimation and some others to an underestimation of the sediment yield at the basin outlets, which has as a favourable consequence the compensation of the deviations between computations and estimates or measurements.

REFERENCES

Beasley, D.B., Huggins, L.F. & Monke, E.J. 1980. ANSWERS: a model for watershed planning. *Transactions of the American Society of Agricultural Engineers* 23(4): 938–944.
Brune, G.M. 1953. Trap efficiency of reservoirs. *Transactions, American Geophysical Union* 34(3): 407–418.

Doorenbos, J. & Pruitt, W.O. 1977. Crop water requirements. *FAO Irrigation and Drainage Paper 24 (revised)*, FAO, Rome, Italy.

Foster, G.R. 1982. Modelling the erosion process. In C.T. Haan, H.P. Johnson & D.L. Brakensiek (eds), *Hydrologic Modeling of Small Watersheds*: Chapter 5, 297–380, American Society of Agricultural Engineers Monograph No. 5.

Foster, G.R., Meyer, L.D. & Onstad, C.A. 1977. A runoff erosivity factor and variable slope length exponents for soil loss estimates. *Transactions of the American Society of Agricultural Engineers* 20(4): 683–687.

Giakoumakis, S., Tsakiris, G. & Efremides, D. 1991. On the rainfall-runoff modelling in a Mediterranean island environment. In *Advances in Water Resources Technology*. Rotterdam: Balkema.

Hrissanthou, V. 1989. Feststofflieferungsmodell eines Einzugsgebietes. *Wasserwirtschaft* 79(4): 186–192.

Hrissanthou, V. 1990. Application of a sediment routing model to a Middle European watershed. *Water Resources Bulletin* 26(5): 801–810.

Hrissanthou, V. 2006. Comparative application of two mathematical models to predict sedimentation in Yermasoyia Reservoir, Cyprus. *Hydrological Processes* 20(18): 3939–3952.

Lutz, W. 1984. Berechnung von Hochwasserabflüssen unter Anwendung von Gebietskenngrößen. *Mitteilungen des Instituts für Hydrologie und Wasserwirtschaft, Universität Karlsruhe*, Germany, Heft 24.

Schmidt, J. 1992. Predicting the sediment yield from agricultural land using a new soil erosion model. In P. Larsen & N. Eisenhauer (eds), *Proceedings 5th International Symposium on River Sedimentation, Karlsruhe*, Germany, 1045–1051.

Schröder, W. & Theune, C. 1984. Feststoffabtrag und Stauraumverlandung in Mitteleuropa. *Wasserwirtschaft* 74(7/8): 374–379.

Schwertmann, U. 1981. Die Vorausschätzung des Bodenabtrages durch Wasser in Bayern. *Institut für Bodenkunde, TU München, Weihenstephan*, Germany.

SCS (Soil Conservation Service) 1972. *National Engineering Handbook*. Section 4: Hydrology, USDA, SCS, Washington DC.

Wischmeier, W.H. & Smith, D.D. 1978. Predicting rainfall erosion losses, A guide to conservation planning. *US Department of Agriculture, Agriculture Handbook No. 537*.

Yang, C.T. & Stall, J.B. 1976. Applicability of unit stream power equation. *Journal of the Hydraulics Division, ASCE* 102(5): 559–568.

Reservoir Sedimentation – Schleiss et al. (Eds)
© 2014 Taylor & Francis Group, London, ISBN 978-1-138-02675-9

The impact of flow transformations and reservoir floor topography on reservoir deposition patterns in high energy environments

F. Weirich

IIHR Hydroscience and Engineering, University of Iowa, Iowa, USA
Department of Earth and Environmental Sciences, University of Iowa, Iowa, USA

ABSTRACT: The challenge of operating reservoirs that are subject to high levels of sediment input or high volume episodic events involves the need to quickly both restore reservoir capacity and regain operational control of outlets works. Such efforts often must be undertaken in the face of increased broader environmental concerns and requirements. In the study presented here the bed configuration of a reservoir resulted in the conversion of a subaqueous debris flow into a turbidity current that relatively quickly dissipated and deposited its sediment load in the upper-mid reaches of the reservoir thereby greatly reducing the volume of sediment reaching the dams intake structures. The results suggest that it may be possible to either alter or take advantage of the reservoir bed configuration to limit the deposition of a significant portion of the sediment to the upper reaches of a reservoir. Such an operational approach might also reduce clean out costs as well as reducing the environmental impact of sediment removal efforts by allowing reservoirs to be only partially drained to remove sediment that has been constrained to deposit in the upper sections.

1 INTRODUCTION

The overall scale and significance of the problem of reservoir sedimentation has been documented in a number of studies such as Mahmood (1987), Yoon (1992) and Shen (1999). From such studies and reviews it is clear that the loss of annual storage capacity varies widely ranging from 0.3% to 2.3% for different regions and local situations. The most commonly agreed upon number suggests that the loss of reservoir storage capacity at a global level averages approximately 1% per year and therefore reservoir sedimentation can present very significant challenges in terms of reservoir management. Moreover, in some situations and regions the challenge of dealing with reservoir sedimentation can be much greater than in others. In settings that are subject to: greater rates of sediment input than the averages cited above; wide variations in rates of sediment influx; or large amounts of sediment being delivered over short periods of time. Also, due to a variety or combinations of factors, large amounts of sediment can accumulate at the outlet works, the problem can be of much greater significance. In particular, the challenge of operating reservoirs that are subject to high rates of sediment or intense high volume episodic sediment input events may require an agency to both quickly restore reservoir storage capacity before another high sediment volume input/flood event occurs and also ensure the rapid restoration and maintenance of operational control of outlet works that may become overwhelmed and left inoperative by such high rate or high volume episodic deliveries of sediment.

In many cases where such high rate or high volume episodic situations exist it has become an even greater challenge in the face of increased environmental concerns and the tendency to expand the role of reservoirs from serving as single purpose facilities, such as flood control, to multipurpose facilities also serving as water supply sources, recreational facilities, and habitat for fish and game. As a result, sediment removal, even in an emergency response to major sediment influx events, is more difficult since conventional mechanical cleanout methods, especially in proximity to the dam face and intakes to the outlet works, often requires the complete draining of the reservoir and/or other relatively high cost methods of sediment removal. In effect, dealing with the impact of reservoir sedimentation, especially in higher rate or episodic high volume events involves two somewhat distinct yet related issues: 1) the actual volume of sediment entering the reservoir that can very rapidly reduces storage capacity and; 2) the distribution of the sediment in the reservoir and in particular the often large amount of sediment reaching the intake areas in such locations or situations and potentially impacting operational control.

These issues are of particular significance for many reservoirs in Southern California, USA where highly erosive steep slopes, relatively steep channel gradients, and sequences of often intensive storm events can generate relatively high sediment load flows into the reservoirs. In addition, many of the drainage areas providing input to reservoirs in Southern California are subject to fire related high volume debris flows that can also reach the reservoirs. Such events often result in large volumes of sediment being transported into and deposited in the reservoirs in relatively short, intense sediment influx events. Of particular concern in this environment, given the high energy nature of many of the flows, is the likelihood that much of the sediment entering the reservoirs can reach the lower portions of the reservoir in the areas of the intake structures, either as turbidity currents, hyperconcentrated flows or in many instances as subaqueous debris flows. In the past, in many situations, one approach to limiting the amount of sediment reaching the outlet works has involved maintaining relatively high water levels in such reservoirs in an effort to have the incoming sediment deposit in the areas where streams enter the upper portion of a reservoir and as far away from the actual dam as is feasible. But such an approach, while potentially effective for incoming flows with lower sediment levels, may not prove effective for the higher concentration turbidity, hyperconcentrated flows and subaqueous debris flows. In this paper the results of an effort to evaluate an alternative approach utilizing reservoir floor topography and flow transformations to alter the deposition pattern of sediment being delivered to a reservoir by such higher concentration flows are presented.

2 BACKGROUND

The occurrence of density flows primarily driven by the presence of sediment was recognized as early as 1885 when Forel (1885), described turbid density flows of the Rhine and Rhone entering Lake Constance and Lake Geneva. Interest in such processes was greatly increased when the occurrence of such sediment driven gravity flows was documented in reservoir settings beginning around the 1920's and with the recognition of such flows from the Colorado River traversing the floor of Lake Mead and reaching the base of the Hoover Dam (Grover & Howard, 1938). Subsequently, Bell (1942), sought to both better document and determine the conditions under which such flows would occur and evaluate their behavior under differing conditions. Subsequently Zhang et al (1976), Sloff (1991) and others have worked to further our understanding of the nature and behavior of turbid density flows from the perspective of lake and reservoir sedimentation dynamics. Much of the effort in this area has also been directed at developing strategies that allow for either controlling the path of such sediment laden flows in an effort to either route such flows through a reservoir, around a reservoir via bypass structures of some type, or even entrain sediment already deposited on the floor of a reservoir and carry such material out of the reservoir via various outlet structures. Morris and Fan (1998/2010), and White (2005) provide summaries of these efforts and strategies.

One aspect of the nature and behavior of such sediment driven gravity flows that can be of significance to reservoir sedimentation processes, especially in those environments subject to high concentration turbidity currents, hyperconcentrated flows and debris flows, is the role of reservoir floor topography and flow transformations on the behavior of such flows. Middleton and Hampton (1976), Parker & Coleman (1986), Middleton (1993), and Iverson, (1997), provide a broad overview of the basic fluid dynamics and deposits of such flows. Hampton (1972), Fisher (1983), Pierson & Scott (1985), Weirich (1988) (1989), Garcia (1993),

Waltham (2004), Felix & Peakall (2006), Kostic & Parker (2006), Felix et al (2009), Spinewine et al (2009) focused on aspects of the question of flow transformations and the mechanisms, processes and circumstances under which subaerial or subaqueous debris or hyperconcentrated flows might convert to turbidity currents once they enter a body of water. The impact of such flow transformations on reservoir sedimentation processes has been evaluated in studies such as Oehy & Schleiss (2007).

3   THE PRESENT STUDY

As part of a wider effort to further our understanding of sediment production, flow and depositional processes in high energy mountain environments, a series of operational experiments were conducted. One aspect of these experiments, which is presented here, specifically focused on the impact of reservoir floor topography on flow transforms of subaerial and subaqueous debris flows into turbidity flow and the potential impact of such processes on reservoir deposition and reservoir sediment management practices and procedures in such environments. This operational experiment involved: 1) instrumenting a watershed feeding a reservoir to monitor flow and sediment levels over a series of storm events; 2) pre event and post event experiment mapping of the bed topography and evaluation of the deposits; 3) the deployment of a 3D array of sensors in the reservoir itself to monitor and track the incoming flows; and 4) the prescribed burning of the instrumented watershed, in part, to increase the likelihood of higher concentration flows during subsequent storm events.

The site selected for this aspect of the study was the San Dimas Reservoir in the San Gabriel Mountains of Southern California (Fig. 1). It is operated as a flood control reservoir

Figure 1.   Location of the experiment, the source sub-basins, instrumentation and flow route of flows to the San Dimas Reservoir. The general location of the subaqueous hydraulic jump discussed below is also indicated by the * (after Weirich (1989)).

by the Los Angeles County Department of Public Works/Los Angeles County Flood Control District (LACDPW/LACFCD). It has a capacity of approx. 1,900,000 m³ and receives drainage from approx. a 42 km² area. Much of the drainage area feeding the reservoir is within the boundaries of the San Dimas Experimental Forest under the jurisdiction of the US Forest Service. As a result of a combination of a relatively steep drainages with slopes averaging above 55% in the majority of the San Dimas watershed, the steep gradient streams with little intermediate storage capacity, relatively highly erodible soils associated with rapid uplift of the San Gabriel Mountains, and a Mediterranean climate pattern of often intense winter storm sequences, often accompanied by high intensity rainfall events, the reservoir is subject to large and episodic influxes of sediment (Weirich, 1989). The frequency of fire in the watershed has resulted in even greater levels of sediment delivery to the reservoir.

4 THE EXPERIMENTAL PROCEDURE AND RECORD OF EVENTS

In order to increase the likelihood of flow events and as part of a wider effort to better understand erosion and sediment transport processes in fire impacted landscapes such as that of Southern California, a series of prescribed burns were carried out. These burns, undertaken in cooperation with the US Forest Services were conducted on selected sub-basins within the San Dimas Experimental forest and which drain into the San Dimas Reservoir. Prior to these burns concrete flumes located at the base of the selected sub-basins, labeled 1,2 and 3 in Figure 1, were instrumented with discharge measuring systems an array of sensors designed to continuously measure the sediment concentration of the flows passing through the flumes during runoff events during runoff events. These sub-basins ranged in size from 160,000 m² to 360,000 m². At the same time an array of sensors were also placed in the reservoir to monitor the passage of flows along the floor of the reservoir. These reservoir sensors were intended to measure the velocity, density and depth of the flows moving along the reservoir floor during runoff events. Prior to the prescribed burn, during the summer of 1984, the reservoir floor was drained and survey data documenting the floor topography of the reservoir was also obtained. In the summer of 1984 sub-basins 1 and 2 (Fig. 1) were prescribed burned. Sub-basin 3 was instrumented but not burnt and was used as a control. Channel surveys from the areas of the flumes down the reservoir were also undertaken.

From November 1984 to May 1985, the monitoring systems tracked the flows and sediment movement. During this period some 19 rainfall events took place that resulted in some level of debris flow event taking place in the burnt watersheds. The debris flow event that took place on Dec. 19, 1984 is of particular interest for our purposes. After a two day period of rain that produced approx. 13 cm of rainfall, a peak rainfall event with an intensity of some 5 mm/5 min occurred at approx. 16:15 hrs. It triggered two almost simultaneous debris flows (they were only 3 minutes apart) in both sub-basins. The debris flows, which last for approx. 15 minutes, were both witnessed by the author.

The sediment concentration record of these events were recorded by the sensors in the flumes at the entrance of sub-basin1 and 2. The record for sub-basin 2 is presented in Figure 2. After passing through the monitoring flumes the two merged debris flows then flowed down toward the San Dimas Reservoir via the main channel of the drainage indicated in Figure 1. Direct observation of the flows indicated they consisted of: a) leading edge or head containing both large and small bolders, cobbles and smaller materials along with large amounts of organic debris of various sizes ranging from logs on downward which lasted only a few minutes; b) a main body of flow consisting of a rather more fluid like flow; and c) a tailing off portion of sediment laden flow. The sensor density data was consistent with the visual observation data. The sensor data also indicated the sediment load was up to approx. 60% sediment by weight for both flows and the total volume of sediment conveyed by these two events was approx. 6,000 m³. Subsequent channel surveys from the flume areas down to the reservoir indicated that approx. 50% of the total volume of sediment from these two, essentially simultaneous, flows was deposited in the channel area. The remainder entered the reservoir. There were no significant flow events following this event.

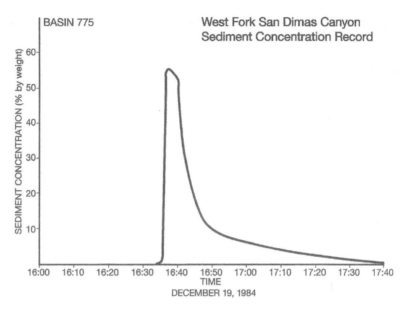

Figure 2. Density data for debris flow event in sub-basin 1 (Based on data from Weirich (1989)).

The subaerial debris flow entered the reservoir and flowed subaqueously across the reservoir floor. Figure 3 shows the condition and water level of the reservoir at the time the debris flows entered the San Dimas Reservoir.

The following summer the reservoir was lowered exposing the majority of the floor area.

Figure 4 shows the exposed floor of the reservoir. The image was taken in June of 1985 following the debris flow event. A month long program of mapping deposits, collection and evaluation of sediment deposits and data from the reservoir sensors was then undertaken.

The mapping and analysis indicated both flows had followed a main channel already present on the floor for several hundred meters before fanning out to cover almost the entire width of the reservoir floor. The thickness and character of the deposits underwent a rather abrupt change after a bend in the channel floor which is visible in Figure 5. Of particular note was the abrupt thickening of the deposits immediately below the bend from a mean thickness of approx. 0.3 m to over 0.5 m. It was a consistent change apparent in a series of cross-sectional profiles made from the reservoir entrance area to well below the bend. The presence of fire related ash and burnt organic materials incorporated into the subaerial and subaqueous debris flows as well as deposits downstream of the bend enabled the recognition of the deposits that were distinctly part of the fire related debris flow documented by the flume sensors. The upper limit of fire ash along the channel walls and on the reservoir floor also allowed for the determination of the actual depth of flow of the subaqueous debris flow, as distinct from the depth of the deposits left by the flow.

The nature of the deposits left by the subaqueous debris flow in the area above the channel bend are indicated in Figure 6a and Figure 6b. The deposits below the bend were more indicative of turbidity current deposits.

The abrupt change in the thickness and nature of the deposits as well as the actual thickness of the subaqueous flows that moved down the channel in the reservoir floor in the area up to and just before the bend in the channel and then downstream of the bend in the channel floor strongly support the interpretation that the subaqueous debris flow had undergone a hydraulic jump. As a result of the jump the subaqueous debris underwent a flow conversion to a turbidity current. Calculation of velocities suggest a pre-jump velocity of approx. 4 m/s were reduced to approx. 1 m/s with a thickening of the flow from approx. 0.5 m to approx. 2.0 m. This was followed by a rapid increase in the

85

Figure 3.   The condition of San Dimas Reservoir at time of the debris flow event.

Figure 4.   The San Dimas Reservoir floor after the debris flow event and the subsequent lowering of reservoir water level to enable the evaluation of the deposition pattern and deposits. Note the channel along the floor extending from the fore-delta area in the foreground down reservoir towards the central section of the reservoir.

deposition rate of material on the reservoir floor. Figure 7 is a mapping of the deposition pattern of the debris flow/turbidity current on the reservoir floor. In the end approx. 70% of the 3,000 m^3 of sediment that entered the reservoir as part of the event was deposited within 200 m of the jump area and within 200 m of the entry point of the flow into the reservoir.

   In this instance the nature of the floor topography appears to have caused a flow conversion that resulted in the rapid deposition of much of the load being transported into the reservoir by a debris flow. As a result sediment that might otherwise been transported the

Figure 5. Arial view of drained reservoir showing area of channel bend and flow conversion zone which is also indicated in Figure 1. The depth of water in the areas of the bend at the time of the debris flow event was on the order of 7 m.

Figure 6a. The deposits of the subaqueous debris flow in the channel on the reservoir floor in the area between fore-delta and area of subaqueous hydraulic jump.

87

Figure 6b.　Close up of some of the deposits in the area shown above.

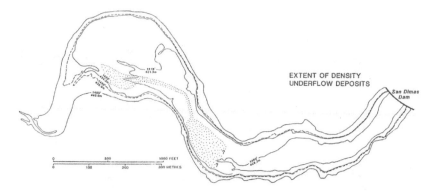

Figure 7.　Extent of deposits from subaqueous debris flow and majority of turbidity current materials. (after Weirich (1989).

full length of the reservoir down to the outlet works was instead emplaced much closer to the inlet where removal might be undertaken at a lower cost and with less environmental issues than if full drainage and excavation around the outlet works would require.

5　CONCLUSIONS

The results clearly demonstrate that reservoir bed topography can significantly, and in some instances, dramatically alter the reservoir sediment deposition pattern. In at least one case the bed configuration resulted in the conversion of a subaqueous debris flow into a turbidity current. The turbidity current then relatively quickly dissipated and deposited its sediment load in the upper-mid reaches of the reservoir thereby greatly reducing the volume of sediment reaching the dams intake structures. The results suggest that it may be possible to either alter or take advantage of the reservoir bed configuration to limit the deposition of a significant portion of the sediment to the upper reaches of a reservoir. Such an operational approach might also reduce clean out costs as well as reducing the environmental impact of sediment removal efforts by allowing reservoirs to be only partially drained to remove sediment that has been constrained to deposit in the upper sections.

REFERENCES

Bell, H.S. 1942. Stratified flow in reservoirs and its use in preventing of siltation. USDA, Msc. Publ. 491, USGPO, Washington, D.C.

Fan, J. & Morris, G.L. 1992. Reservoir sedimentation. II. Desiltation and long-term storage capacity. *J. of Hydraulic Eng.*, 118(3):370–384.

Felix, M., & Peakall, J. 2006. Transformations of debris flows into turbidity current: mechanisms inferred from laboratory experiments. *Sedimentology*, 53:107–123.

Felix, M., Leszcynski, S., Slaczka, A., Uchman, A., Amy, L., Peakall, J. 2009. Field expressions of the transformation of debris flow into turbidity currents, with examples from the Polish Carpathians and the French Alps, *Marine and Petroleum Geology*, 26:2011–2030.

Forel, F.A. 1885. "Les Ravins, Sous-lacustres de Fleuves Glaciares" *Comptes Rendus, Academie de Sciences*, Paris, T. 101:725–728.

Fisher, R.V. 1983. Flow tansformations in sediment gravity flows. *Geology*, 11:273–274.

Garcia, M.H. 1993. Hydraulic jumps in sediment-driven bottom currents. *J. of Hydraulic Eng,.* 119:1094–1117.

Grover, N.C. & Howard, C.S. 1938. The passage of turbid water through Lake Mead. *ASCE*, 103:720–790.

Hampton, M.A.1972. The role of subaqueous debris flow in generating turbidity currents *Journal of Sedimentary Petrology*, 42:775–793.

Iverson, R. 1997. The physics of debris flows. *Rev. Geophys. Res.* 35:245–296.

Kostic, S. & Parker, G. 2006. The response of turbidity currents to a canyon-fan transition: internal hydraulic jumps and depositional signatures, *J. of Hydraulic Res.*, 44:5, 631–653.

Mahmood, K. 1987. Reservoir Sedimentation: Impact, extent, and Mitigation. World Bank Technical Paper No. 71, The International Bank for Reconstruction and Development.

Middleton, G.V. & Hampton, M.A. (1976) Subaqueous sediment transport and deposition by sediment gravity flows, in Stanley, D.J. and Swift, D.J.P. eds., *Marine Sediment Transport and Environmental Management*: New York, Wiley. p197–218.

Middleton, G.V. 1993. Sediment deposition from turbidity currents, *Annu. Rev. Earth Planet. Sci.*, 21:89–114.

Morris, G.L. & Fan, J. 1998. *Reservoir Sedimentation Handbook.*, New York: McGraw-Hill.

Morris, G.L. & Fan, J. 2010. *Reservoir Sedimentation Handbook: Design and Management of Dams Reservoirs, and Watershed for Sustainable use (electronic version)*. New York: McGraw-Hill.

Oehy, C. & Schleiss, A. 2007. Control of turbidity currents in reservoirs by solid and permeable obstacles, *J. of Hydraulic Eng.*, 133(6):637–648.

Parker, G. & Coleman, N.L. 1986. Simple model of sediment-laden flows, *J. Hydraulic Eng.* 112:356–375.

Pierson, T.C. & Scott, K.M. 1985. Downstream dilution of a lahar transition from debris flow to hyperconcentrated flow: *Water Resources Research*, 21:1511–1524.

Sloff, C.F. 1991. Reservoir sedimentation: A Literature Survey. Report No. 91-2. Communications on Hydraulic and Geotechnical Engineering, Delft University of Technology, Fac.of Civil Eng, 124p.

Shen, H.W. 1999. Flushing of sediment through reservoirs. *J. of Hydraulic Res.*, 37(6):743–757.

Spinewine, B., Sequeiros, O.E., Garcia, M.H., Beaubouef, R.T., Sun, T., Savoye, B., & Parker, G, Experiments on wedge-shaped deep sea sedimentary deposits in minibasins and/or channel levees emplaced by turbidity currents. Part II. Morphodynamic evolution of the wedge and of the associated bedforms, *Journal of Sedimentary Research*, 79:608–628.

Waltham, D. 2004. Flow transformations in particulate gravity flows. *J. of Sed. Res.*, 74:129–134.

Weirich, F. 1988. Field evidence for hydraulic jumps in subaqueous sediment gravity flows. *Nature*, 332:626–629.

Weirich, F. 1989. The generation of turbidity currents by subaerial debris flows, California. *Geological Society of America Bulletin*, 101:278–291.

White, W.R. 2005. A review of Current Knowledge World Wide Water Storage in Man-Made Reservoirs. FR/R0012, Foundation for Water Research Allen House, Liston Road, Marlow. pp 40.

Zhang, H., Xia, M., Chen,S.-J., Li, Z-W., & Xia, H.-B.,1976. Regulation of sediments in some medium and small-sized reservoirs on heavily silt-laden streams in China. *12th ICOLD*, Q.47, R32: 1223–1243.

Reservoir Sedimentation – Schleiss et al. (Eds)
© 2014 *Taylor & Francis Group, London, ISBN 978-1-138-02675-9*

Reservoir sedimentation and erosion processes in a snow-influenced basin in Southern Spain

A. Millares
Andalusian Institute for Earth System Research, University of Granada, Spain

M.J. Polo
Andalusian Institute for Earth System Research, University of Córdoba, Spain

A. Moñino, J. Herrero & M.A. Losada
Andalusian Institute for Earth System Research, University of Granada, Spain

ABSTRACT: Soil erosion is one of the major problems of Mediterranean watersheds, being reservoir sedimentation and silting the final consequence of the dynamics of erosion-transport-deposition processes upstream. In mountainous areas, the influence of the snowmelt/accumulation cycles on the water flow regime greatly determines the river configuration and conditions fluvial dynamics, causing erosion and significant loads of sediment. This work presents the results from the monitoring of reservoir silting in a mountainous coastal watershed in Southern Spain, where the predominant erosion processes and their significant thresholds have been identified, together with the effect of the sediment retention dams located along the river. The Rules reservoir in the Guadalfeo River Basin was finished in 2004, closing a draining network of approximately 1000 km^2 and providing a maximum storage of approximately 110 hm^3 for urban and agricultural demands in the area. This coastal basin includes the Sierra Nevada mountain range, with the highest altitude in the Iberian Peninsula, up to more than 3000 m, and the coastline only 40-km away. This extreme topographic gradient has produced a wide range of climate environments, soils, vegetation, and morphologic features throughout the basin. The 8-yr monitoring of the basin since the reservoir's construction has demostrated the inefficacy of the sediment retention structures upstream to prevent silting processes, and the significance of both torrential rainfall and intense snowmelt events for generating huge sediment pulses. The monitored siltation of an in-river check-dam located upstream, with 0.2 hm^3 of sediment retention in just 6 years associated with 2 intense events, together with the capacity loss of the Rules reservoir since 2008, confirmed the order of magnitude of the erosion processes and the importance of bedload in this semi-arid mountainous environment. Moreover, the sediment distribution accumulated along the reservoir allowed the identification of sediment sources and the partition between fluvial and hillslope erosion processes. The total sediment production is estimated at 24 $t \cdot ha^{-1}y^{-1}$ with fluvial contributions of 55% of the partitioning between suspended and fluvial loads. However, the great spatial and temporal variability of the processes involved reduces the validity of simple annual average approaches.

1   INTRODUCTION

Quantification of both soil loss and fluvial transport and sedimentation processes is essential for a proper management and conservation of our territory. This is especially true in semi-arid environments, where the huge amount of available sediment at the river floodplains and the uncertainty and intensity of precipitation events lead to important damage along the rivers and hillslopes, and a severe reservoir siltation (Graf & Lecce 1988, Coppus & Imeson 2002). In these environments, quantification based on a single datum of specific sediment yield, SSY, (e.g.; 10 $t \cdot ha^{-1} \cdot y^{-1}$) is not representative due to the great spatial variability of erosion processes and the high temporality of the forcing agents. From this holistic perspective, both predictive models and measurement methodologies face significant challenges.

The quantification based on hillslope processes such as the Universal Soil Loss Equation (USLE or its revised version RUSLE) (Wischmeier, 1978) or on monitoring the volume loss in hillslope check-dams, do not consider fluvial processes such as bedload. Although bedload is generally ignored for estimating SSY, it is essential to quantify the total sediment delivery of basins. These contributions could exceed over 50% of the fraction between suspended load and bedload (Schick & Lekach 1993, Turowsky et al. 2010), especially in semiarid environments (Graf & Lecce 1988, Powell et al. 2001, Millares et al. 2014). Although reservoir sedimentation is a serious problem which reflects the consequences of soil erosion, it also provides an opportunity for understanding erosion and sedimentation transport processes at a basin scale. Based on the accumulated volume in medium to large dams, it is possible not only to estimate the total production of sediment but also to distinguish between erosion processes by assuming a natural separation of sediment along the dam. This hypothesis consists of the assumption of a suspended sediment and bedload separation (submerged area and tail) along the body of the dam, as reported in other field studies (Duck & McManus 1994, Rowan et al. 1995, Snyder et al. 2004, Minear et al. 2009).

The main goal of this research was to assess the variability of sediment yield in a semi-arid basin, taking into account both fluvial and hillslope processes by monitoring the silting evolution of two control volumes. Surveys during the period 2007–2012 in the Rules Reservoir (110 hm^3), in southern Spain, have permitted the quantification of the sediment yield of the basin and the discretization between fluvial and hillslope deposits. Bathymetric, topographic and analytical characteristics of the sediment carried out in the reservoir were complemented by measures of both suspended load and bedload along the river (Millares et al. 2014). The validation of these river measurements from the two control volume selected, allowed us to understand the erosion dynamics of the basin and to delimit the spatial distribution of the dominant erosion processes.

2 STUDY SITE

The study area comprises a significant part, 650 km^2, of the headwaters of the Guadalfeo river (1250 km^2) located in the south-east of the Iberian Peninsula (Fig. 1). The mountainous influence of Sierra Nevada, with 3780 m. asl., conditions the hydrologic dynamics and the pluvio-nival character of this semi-arid basin. The annual precipitation is marked by a great spatio-temporal variability, important altitudinal gradients and the occurrence of extreme events that may exceed 300 mmd^{-1} (Capel 1974).

From a geologic and geomorphologic point of view, three units can be distinguished with different hillslope erosion dynamics (see Fig. 1):

1. The south side of Sierra Nevada massif with large altitudinal gradient (300–3450 m) and a moderate erosion range (10–25 $t \cdot ha^{-1} \cdot yr^{-1}$), related to rill and interrill erosion processes (Ministerio de Medio Ambiente Rural y Marino 2007).
2. The Sierra de la Contraviesa with a more advanced geomorphology, presents a greater variety of hillslope processes with frequent gullying and mass movements processes. The soil loss associated with this area could, according to the RUSLE model, exceed 200 $t \cdot ha^{-1} \cdot y^{-1}$ (Ministerio de Medio Ambiente Rural y Marino 2007). Periodic severe events lead to important bedload contributions that accumulate in the main channel of the Guadalfeo river (Millares et al. 2014).
3. The Sierra de Lújar, despite its rocky nature and poorly developed soils caused by recurrent wildfires, presents ensigns of gullying processes associated with non-terraced almond crops. Soil loss is quantified in 10–20 $t \cdot ha^{-1} \cdot y^{-1}$ estimated from RUSLE model (Ministerio de Medio Ambiente Rural y Marino 2007).

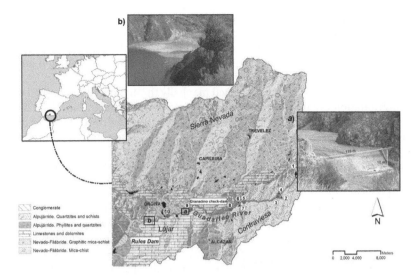

Figure 1. Location of the study basin and the locations selected for monitoring suspended and bed-load sediment. Points 1 to 10 show the sample location for suspended sediment campaigns in the main river (red circles) or at the sub-basin outlet (yellow triangles). Squares *a* and *b* show the two control volumes monitored: Granadino check-dam (a) and Rules reservoir (b). The geologic differences between the three areas described: Sierra Nevada (north), Sierra de la Contraviesa (south-east) and Sierra de Lújar can be observed.

The wide alluvial floodplain, which forms the main channel of the river, with section lengths ranging from 30–140 m and an average river slope of 0.011 $m \cdot m^{-1}$, is comprised of Neogene-Quaternary deposits. The river has an essentially straight flow pattern although there are some minor derivations. The nival contributions condition the quasi-perennial flow, which allows the development of an armor layer and separates a surface layer ($d_{50} \approx 60$ mm) from a substrate ($d_{50} \approx 2.5$ mm). The periodic occurrence of intense precipitation and snowmelt events, with flow rates that can exceed 1000 $m^3 s^{-1}$ reshape this drainage network and release a large amount of sediment. In this regard, (Millares et al. 2014) reported the successive stages observed at the ephemeral channels of Sierra de la Contraviesa of the accumulation/erosion pattern of this area. These contributions are first accumulated as alluvial fans at the respective connections to the main channel, and are later eroded and transported along the Guadalfeo River up to the Rules reservoir. The sediment volume measured during 6 years at the headwater part of the basin (482 km^2) pointed to bedload contributions of up to 600 $t \cdot km^{-2}$ related to an intense event in 2009 with an averaged value of 100 $t \cdot km^{-2} \cdot y^{-1}$ for the period of 2004 to 2010. This result indicates the fluvial nature of the sediment contributions at the study site and points to an analysis based on the partitioning of fluvial and hillslope processes.

3 METHODS

The work developed in this study during the period 2004–2011 includes three different approaches;

- The monitoring of suspended sediment through 10 control points located both within the main river and at the outlet of ephemeral rivers.
- The river bed configuration and the bedload study by monitoring the accumulated sediment at the Granadino check-dam.
- Monitoring of accumulated sediment in the Rules reservoir by bathymetric and topographic surveying in both emerged and submerged areas respectively.

In order to quantify the forcing agent regime, three new weather stations, in addition to the eight existing ones, were installed in the study area. The fraction of precipitation falling as snow and the contributions of snowmelt were estimated with the physically based hydrological model WiMMeD (Herrero et al. 2010), which was calibrated and validated for the study area.

The suspended load samples were collected manually throughout different field campaigns for both precipitation events and periods of snowmelt from April to June. The distribution of these points, as shown in Figure 1, was selected in order to identify differences in sediment production related to the geologic and geomorphologic settings described in the previous section. At these points a total of 204 samples were collected. To determine the suspended sediment concentration mgl^{-1}, the samples obtained were analysed by Whatman GF-F filters of 0.7 μm pore size, following the TSS 160.2 protocol (US Environmental Protection Agency 1999) (see Fig. 2a).

The bed material of the main river was characterized from data obtained during the summer months. Surface and substrate samples were obtained and the size distribution for each was determined using sieve analysis (Fig. 2b). The dry specific mass corresponding to the samples, ρ_{md}, provided a mean value of 1453 kgm^{-3} with $\sigma = 133$ kgm^{-3}, the value of which depended on the depth and distribution of the sediment diameter. The sediment accumulation in the Granadino check-dam was monitored by 11 field surveys between 9 February 2004 and 27 August 2010 (Fig. 2c). Details of this works can be found in (Millares et al. 2014).

The bathymetric works in the submerged area of the Rules dam consisted of two surveys with an Imagenex multibeam echosounder with data correction for heading, depth, pitch, heave and roll. The positional data were acquired with a Trimble Ag-132 GPS, with accuracy of 2 cm. The final resolution of the generated surfaces was 10×10 m. The topographic works in the delta (emerged area) were performed using the differential GPS generating surfaces of

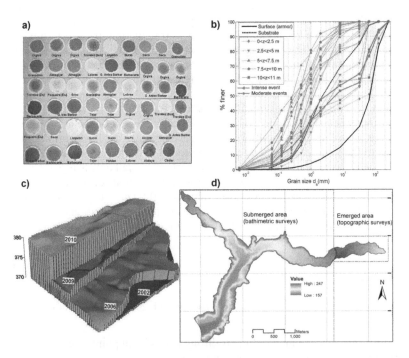

Figure 2.   Suspended sediment samples at different locations along the river, showing differences in colour and mineralogic composition of the contributing sub-basins (a). Results corresponding to grain size distributions of registered sediment at the Granadino check-dam according to depth intervals (0.5 m) and average of the bed river layers (b). Results of topographic campaigns and generated surfaces at the check-dam for different years (c). Bathymetric and topographyc surfaces at Rules reservoir (d).

10×10 m of resolution (Fig. 2d). Although the reservoir was completed in 2003, work on adapting the terrain was carried out until 2007, when its functioning began.

4   RESULTS AND DISCUSSION

The data measured of bedload and suspended sediments were compared with the accumulated volume at the Rules dam throughout the study period. Figure 3 shows the relationship between the flow and the measured suspended sediment at two output points located at the main river; Granadino and Órgiva. As can be observed, there is a potential behavior $(SS = aQ^b)$ which can be compared with previous works carried out in Mediterranean environments (e.g. $10 < a < 1340$, $0.5 < b < 2$; Farguell & Sala 2006). Furthermore, the relationship between suspended sediment concentration and discharge of the 10 control points selected shows an important contribution of the hillslope erosion processes corresponding to Sierra de la Contraviesa (points 4 to 8 in Fig. 1) strongly related to intense events. Along these sub-basins, during the medium-intensity event of 2005, concentrations of up to 3500 mgl^{-1} were measured at the outlet with a total rate estimated from the maximum flow and the event duration between 0.23–0.95 $t \cdot ha^{-1} event^{-1}$.

The fitted curve, together with the analysed characteristics of the suspended loads, allowed us to estimate, approximately, the dynamics of suspended sediment and the associated mass which travelled through the river during the study period. Considering the flow recorded, the total rate of sediment transported by suspended load could be estimated at 13.34 $t \cdot ha^{-1} y^{-1}$ with a total mass of $2.65 \cdot 10^6$ t of soil loss for the entire basin.

Moreover, the accumulated bedload in the two control volumes located at the main river Granadino check-dam and the tail of the Rules reservoir allowed us to quantify these fluvial contributions. The use of these control volumes, despite limitations (lack of data during the event, hydrodynamic effects that rearrange bedload layers, or difficulties in assessing spatial distribution), has proven to be advantageous. These methodologies allow a broad understanding of the transport processes in comparison to others, such as that of Helley-Smith or Birkbeck slot samplers, with limited measurement volumes for intense events.

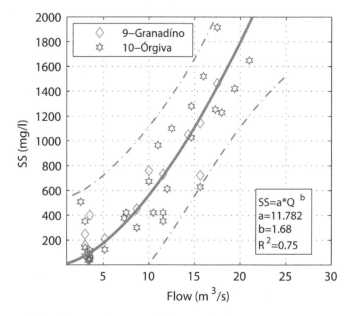

Figure 3.   Suspended sediment rating curves (SS; suspended sediment concentration, Q; water flow) at Granadino and Orgiva control points corresponding to the precipitation events registered in 2004, 2005 and 2009.

The results obtained show a differential behavior in two stretches of the main channel. From the Granadino check-dam to upstream, the existence of an armor layer was seen to greatly control the bedload input. The results, detailed in Millares et al. (2014), suggest a greater transport efficiency than that usually found in mountain rivers of a marked torrential nature. The effective runoff estimated from the critical flow Qc has been assessed as $Q_c = 5.5 \, m^3 s^{-1}$, which is similar to that found in other torrential mountain rivers studies. The bedload measured during moderate and intense rainfall/snowmelt events displayed equivalent proportions of cobble-gravel and sand. This confirms the controlling role of the armored layer in this stretch of the river and points to the predominance of near-equal mobility processes. However, the stretch between the Granadino check-dam and the tail of the Rules reservoir behaved very differently for high-intensity events, resulting in a considerably larger production of sediment. The intense event recorded in December 2009 accumulated approximately 0.143 hm^3 of sediment in the check-dam and approximately 1.80 hm^3 at the tail of the Rules reservoir, this means a higher order of magnitude. This large increase in sediment in this stretch could be related to the contributions from other processes such as bank erosion and avulsion during intense events. However, the existence of a quarry close to the floodplain makes it advisable to also consider an anthropogenic influence to explain this significant increase in sediment production.

Table 1 shows the volumetric results at both control volumes subdividing the emerged and the submerged areas of the Rules Reservoir. As can be observed, the submerged area includes the quantification of quarries used to obtain material (sand and fine gravel) for refilling the nearby beaches and increasing the volume of the dam before the filling process. Also, the artificial sediment contribution corresponding to the pillars of the A-7 highway was assessed. These volumes were discarded for the analysis.

In Figure 4a,b the differences between the obtained surfaces are exposed. The scale represents the height in meters of a column of deposited or eroded sediment with a resolution

Table 1. Volumetric results corresponding to the selected control volumes during different stages of the study period. As can be observed quarries and pillars has been quantified at the Rules reservoir to avoid its influence on the total calculation.

Control volume	Effective period	Submerged area (hm^3)			Emerged area (hm^3)
		External contribution	Quarries	Pillars	
Granadino check-dam	2004–2007	–	–	–	0.0610
	2007–2010	–	–	–	0.1430
	Total				0.2040
Rules reservoir	2008–2011	1.64	–0.7251	0.1832	1.80

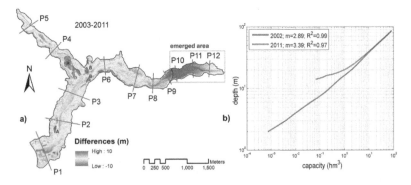

Figure 4. Surface differences calculated from bathymetric and topographic surveys, submerged and emerged area respectively, for the periods of 2003–2011 (a). Locations of quarries (A) and pillars (B) are clearly identified. Reservoir depth-capacity relationship for the three surfaces used (b).

of 10×10 meters. The emerged part can be easily identified (P10 to P12) with a maximum height of 16 meters of deposited sediment, mostly accumulated during the intense event of 2009. The size distribution of this deposited sediment coincides with the bedload size measured at the Granadino check-dam for the same event. It is interesting to observe how, during the following measurement period 2011–2012, there is a significant shift of this material into the submerged area, which does not go far in distance. This implies that, even without events of any great intensity, a significant erosion of the coarser material occurs which reflects a high rate of sediment entrainment.

As can be identified, most of the deposited sediment in the submerged area is located following the old path of the river (Fig. 4a). Also, an internal landslide can be observed (P6 to P8) at the submerged hillslope of the dam. Only in some specific areas do these landslides include emerged material so, for this study, these contributions could be considered to be neglected in respect of the total volume. The total amount of the net external contributions has been estimated at 1.64 hm^3 for the period 2008–2011. This estimation of net loss of volume is subject to errors related to the methodology used in the bathymetric surveys, mainly due to the calibration of the speed of sound and the variability of submerged sediment density. From the analysed samples, an average value of 1200 kgm^{-3} could be selected, which fits in with estimations suggested by other authors (Morris & Fan 1998). With this consideration, the total mass related to suspended sediment accumulated in the submerged area could be estimated to be of $1.969 \cdot 10^6 t$.

The relationship between depth (m) and capacity (hm^3) in a log-log plot is shown in Figure 4b. As can be observed a significant loss of volume is produced in the central reservoir basin. The calculated slope m of the fitted line, which is the reciprocal of the slope (Morris & Fan 1998), points to a transition from foothill (Type II) to lake reservoir shape (Type I) if we consider the classification proposed by (Strand & Pemberton 1987). This morphological transition is conditioned by the numerous submerged landslides.

5 CONCLUSIONS

The monitoring work at two control volumes of a semi-arid mountainous basin has enabled the assessment of the sediment delivery associated with hillslope/fluvial erosion processes. The measured amount of sediment in the submerged part of the Rules reservoir differ with the estimation made from field campaigns of suspended sediment concentration and the measured flow during the study period, i.e. $2.65 \cdot 10^6 t \neq 1.97 \cdot 10^6 t$. Although these estimations are subject to uncertainty related to the accuracy of the fitted flow/suspended sediment relationship and the density of the submerged sediment, the double methodological approach permitted us to estimate an averaged soil loss rate for hillslopes of 11.6 $t \cdot ha^{-1} y^{-1}$. This average value is subject to a great spatial variability, with a significant increase in the Sierra de la Contraviesa. The conjunction of soft and highly degraded hillslopes in this part of the basin, where gullying and landslide processes are important, and the fractured material in the highest northern areas, subject to sudden snowmelt events, ensures a continuous supply of suspended sediment annually.

Moreover, the contributions coming from fluvial processes within the river have shown themselves to be of a greater importance in the total sediment delivery. The total amount was estimated at $2.48 \cdot 10^6 t$ during the study period, which means approximately the 55% of the partitioning between suspended load and fluvial contributions, much higher than the estimations made in previous studies (Turowsky et al. 2010). Here, the fluvial transport processes include not only the bedload contributions but also other processes such as bank erosion and fluvial avulsion, which increase by one order of magnitude the volume of sediment transported. In this case, the fluvial yield could be estimated at 12.49 $t \cdot ha^{-1} y^{-1}$.

With these results, the total production could be estimated at 24 $t \cdot ha^{-1} y^{-1}$ which is higher than the averaged assessment made with RUSLE for the hillslopes of the study site. This reflects a quantification made by RUSLE methodologies as simplistic as inadequate for understanding the erosion dynamics in these environments, where a great spatial and temporal variability of the erosion processes reduces the validity of simple annual average assessments.

Although the volume loss in the reservoir during the study period represents about 3% of its capacity, the magnitude of the responsible event, with maximun discharge of $Q_p = 100 \ m^3 s^{-1}$ and the associated amount of the fluvial contributions, suggests the worst scenario in terms of reservoir service life in a river with records of $Q_p = 1000 \ m^3 s^{-1}$.

REFERENCES

Capel, J. 1974. Génesis de las inundaciones de octubre de 1973 en el sureste de la Península Ibérica. *Cuadernos de Geografía* 4: 149–166.

Coppus, R. & Imeson, A.C. 2002. Extreme events controlling erosion and sediment transport in a semi-arid sub-andean valley. *Earth Surface Processes and Landforms* 27(13): 1365–1375.

Dirección General de Medio Natural y Política Forestal 2007. *Inventario Nacional de Erosión de Suelos. Comunidad Autónoma de Andalucía*. Granada. Madrid: Ministerio de Medio Ambiente Rural y Marino.

Duck, R. & McManus, J. 1994. A long-term estimate of bedload and suspended sediment yield derived from reservoir deposits. *Journal of Hydrology* 159(1–4): 365–373.

Farguell, J. & Sala, M. 2006. Seasonal Trends of Suspended Sediment Concentration in a Mediterranean Basin (Anoia River, NE Spain). In Owens P.N., Collins A.J. (ed.), *Soil Erosion and Sediment Redistribution in River Catchments*. Cambridge: CABI.

Graf, W.L. & Lecce, S.A. 1988. *Fluvial processes in dryland rivers*. New York: Springer-Verlag.

Herrero, J., Aguilar, C., Millares, A., Egüen, M., Carpintero, M., Polo, M.J., & Losada, M.A. 2010. *WiMMed. User Manual v1.1*. Granada: University of Granada.

Millares, A., Polo, M.J., Moñino, A., Herrero J., & Losada, M.A. 2014. Bedload dynamics and associated snowmelt influence in mountainous and semiarid alluvial rivers. *Geomorphology* 206: 330–342.

Minear, J.T. & Kondolf, G.M. 2009. Estimating reservoir sedimentation rates at large spatial and temporal scales: A case study of California. *Water Resources Research* 45: W12502.

Morris, G.L. & Fan, J. 1998. *Reservoir Sedimentation Handbook*. New York: McGraw-Hill.

Powell, D.M., Reid, I., & Laronne, J.B. 2001. Evolution of bed load grain size distribution with increasing flow strength and the effect of flow duration on the caliber of bed load sediment yield in ephemeral gravel bed rivers. *Water Resources Research* 37(5): 1463–1474.

Rowan, J.S., Goodwill, J.P., & Greco, M. 1995. Temporal variability in catchment sediment yield determined from repeated bathymetric surveys: Abbeystead Reservoir, UK *Physics and Chemistry of the Earth* 20(2): 199–206.

Schick, A.P. & Lekach, J. 1993. An evaluation of two ten-year sediment budgets, Nahal Yael, Israel. *Physical Geography* (3): 225–238.

Snyder, N.P., Rubin, D.M., Alpers, C.N., Childs, J.R., Curtis, J.A.,Flint, L.E., & Wright S.A. 2004. Estimating accumulation rates and physical properties of sediment behind a dam: Englebright Lake, Yuba River, northern California. *Water Resources Research* 40(11): W11301.

Strand, R. & Pemberton, E. 1987. Design of small dams. *Reservoir sedimentation*. Washington: US Department of the Interior, Bureau of Reclamation.

Turowski, J.M., Rickenmann, D., & Dadson, S.J. 2010. The partitioning of the total sediment load of a river into suspended load and bedload: a review of empirical data. *Sedimentology* 57(4): 1126–1146.

US Environmental Protection Agency 1999. *Method 160.2: Total Suspended Solids (TSS) (Gravimetric, Dried at 103?105 °C). Revised Ed*. Washington: United States Environmental Protection Agency.

Wischmeier, W.H. & Smith, D.D. 1978. *Predicting rainfall erosion losses. A guide to conservation planning*. Washington: USDA.

Reservoir Sedimentation – Schleiss et al. (Eds)
© *2014 Taylor & Francis Group, London, ISBN 978-1-138-02675-9*

Suspended sediment dynamics of Ribarroja Reservoir (Ebro River, Spain)

M. Arbat-Bofill, E. Bladé, M. Sánchez-Juny, D. Niñerola & J. Dolz
Institut Flumen UPC-CIMNE, Universitat Politècnica de Catalunya—BarcelonaTech, Barcelona, Spain

ABSTRACT: In order to study the sediment dynamics of Ribarroja Reservoir, a bathymetric campaign was carried out in 2007 using a precision multi-beam probe. One year later, another campaign was performed. This second bathymetric campaign focused only in the tail of Ribarroja Reservoir (confluence Ebro-Segre Rivers), it was designed to study suspended sediment entering the reservoir and its bed evolution in one year. Between 2007 and 2008 campaigns, only one major flood episode took place (from late May to early June 2008). Two different digital elevation models of Ribarroja Reservoir were generated. The analysis and comparison of 2007 and 2008 digital elevation models showed that 170.000 m^3 of sediment (initially located immediately after the Segre-Ebro confluence) were resuspended and displaced about 4 km downstream. This study was complemented with a numerical simulation using the suspended sediment transport module implemented in a two-dimensional depth averaged model.

1 INTRODUCTION

The construction of many Spanish reservoirs reached its peak in the sixties. Nowadays the influence of sedimentary processes is being appreciated in many of them because its storage capacity has been reduced. As the twentieth century was characterized by the construction of large dams, a challenge for the twenty-first century is the conservation of existing structures: maintenance, cleaning or dredging in some cases (Morris & Fan Jiahua 1998).

In recent years significant progress has been made in understanding the importance of the factors involved in erosion and deposition of sediments in rivers and/or reservoirs. However, the forecast of sediment dynamics and the accumulation calculus remains complex and difficult to study. There is still a highly uncertainty in the estimation of the space and time varying patterns of sedimentation in reservoirs. This lack of predictability is related to several factors: the variable flow rate of the inflows/outflows, the load of sediment in suspension, the size of the sediment particles, the specific weight and the physical characteristics of the sediment as well as the geometry of the reservoir. Thus, sediment dynamics evolution is site specific and, because of this, it is often recommended to perform specific studies for each environment or study case.

In natural conditions, a river system tends to an equilibrium between inputs and outputs of sediment. The construction of a dam implies the cut off and the slowdown of the flow reaching the reservoir. This process means that the sediment accumulation occurs in a dam. Sometimes, due to the flow deceleration, when a river is approximating to a reservoir, the sediment capacity of transport is reduced and as the flow slows down, the input sediments are deposited along the tail of the reservoir.

A great number of conceptual and empirical models have been developed and applied to estimate annual sedimentation rate in reservoirs, most of them based on field observations. The commonly studied parameters are the average sediment accumulation and the cumulative sediment volume after a certain number of years of operation (Strand & Pemberton 1982; Morris et al. 2008; Morris & Fan Jiahua 1998).

The impacts of the sedimentation processes in reservoirs can be summarized as (Garcia 2008): decline of the reservoir capacity, problems and failures in performance, increase of the turbidity or in the concentration of organic sediments. Upstream, the problem is due to the deposition of sediments. When the velocity decreases (while the depth increases) sediments are deposited. Downstream of the dam, the problem is well known and highly discussed: the presence of the dam reduces the suspended sediment load, which often results in a non-stoppable tendency to downstream river erosion.

In 2003 The World Bank organization launched an ambitious project named "RESCON" where the bases of sediment management theories were established. In terms of management and maintenance of dams and reservoirs the work was summarized in two volumes "Reservoir conservation: The RESCON approach" (Palmieri et al. 2003; Kawashima et al. 2003).

2 STUDY SITE

In the Ebro River basin there are more than 109 reservoirs with greater capacity than 1 hm^3 and about 800,000 ha of irrigated land (Prats et al. 2007). Ribarroja Reservoir is located in the lower Ebro River (41° 180 N, 0°210 E) in the Mequinenza-Ribarroja-Flix system. The upstream reservoir is Mequinenza (1534 hm^3) and the downstream one is Flix (11.4 hm^3). In the tail of Ribarroja Reservoir (the whole system in Fig. 1, and a zoom of the tail in Fig. 2) there is the Segre River mouth. Segre River is one of the main tributaries of Ebro River, which provides almost half of the Ribarroja Reservoir inflow depending on the period of the year.

Ribarroja Dam was finished in 1969 and since then it is used for energy generation, water supply and irrigation in addition to flood control (LIMNOS 1996). Ribarroja Reservoir has an irregular morphology (Fig. 1) due to the surrounding topography, and its maximum volume is estimated to be 210 hm^3. The residence time is about 6 to 10 days.

According to the Ebro River Water Authority (http://www.chebro.es), Ebro River has an average annual contribution of 8,009 hm^3 to Ribarroja Reservoir. Segre River increases this contribution to 14,069 hm^3. This means that Ebro River contributes on average 58.82% of the inflows to Ribarroja Reservoir, while Segre inputs are 41.15% (Prats-Rodríguez et al. 2011). There is also a small stream, Matarraña River, with a very limited flow contribution.

During the considered study period (between years 2007 and 2008) only one major flood episode took place (from late May to early June 2008). The flood episode presented an important peak coming mainly from the Ebro River, in late May—early June of 2008 (Fig. 3). It presented a maximum average daily flow of 1600 m^3/s measured at the outlet of Ribarroja Reservoir, and it exceeded the 500 m^3/s of discharge for 12 days. During the episode no sediment was provided by Ebro River because it is retained in the Mequinenza Reservoir immediately upstream.

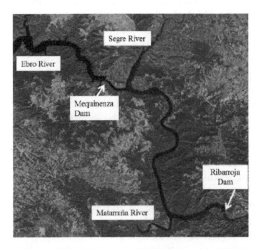

Figure 1.   Location of Mequinenza, Ribarroja and Flix Reservoirs along the Ebro River.
Source: Google Maps, edited by the authors.

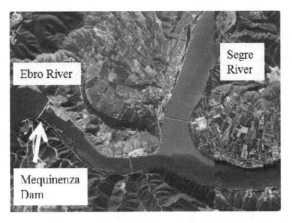

Figure 2. Zoom of the confluence of Ebro River (water exiting Mequinensa Dam) and Segre River (water with suspended sediments).
Source: Google Maps, edited by the authors.

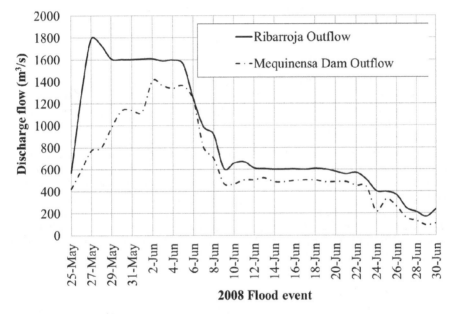

Figure 3. Average daily outflows (m³/s) of Mequinenza (dashed line) and Ribarroja (solid line) Dams.

The sediment of Ribarroja reservoir was characterized in previous studies (López et al. 2012). Throughout almost the whole Reservoir, the sediments mainly exposed a lime-clay texture, with a percentage of lime (4 μm < φ < 63 μm) between 56% and 74%, and a percentage of clay (φ < 4 μm) between 18% and 43%.

3 METHODS

3.1 2007 bathymetric campaign

In order to study the dynamics of Ribarroja Reservoir in detail, an in-site campaign was conducted to obtain an accurate digital model of the bottom of the reservoir. In autumn 2007 a field bathymetric campaign was carried out by using a multi-beam sensor. The sampling of the profiles was performed using two echo sounders (one double-frequency beam and one multi-frequency beam) placed on a boat equipped with a real-time DGPS positioning. SEXTANT hydrography software was used for data collection and also for navigation paths planning. In concordance with the use of topographic databases of Cartographic Institute of Catalonia (ICC), the ED50 UTM coordinate system was used. The gathering of the bathymetric rough data of the reservoir was made by several series of transverse and longitudinal profiles along the course of the river.

According to the demands of field sampling (different requirements of precision for data collection), the area of study was divided into three subsections of approximately 10 Km each (see Fig. 3). Zone 1: from the Mequinenza bridge in Segre River, over 10 km downstream (zone of most interest because of the confluence), detailed measurement with multi-beam probe. Zone 2: From the end of Zone 1 to 10 km downstream, using the single beam probe. The profiles were performed every 25 m along the cross section, transverse to the axis of the river. Zone 3: from the end of Zone 2 to Ribarroja dam (approximately 10 km) longitudinal profiles and cross-section profiles were taken every 100 meters. Besides these three main areas it was considered necessary to study in detail two important zones within Zone 3: the Matarraña River mouth and the vicinity of Ribarroja Dam (cross sections every 10 m).

The echo-sounder was calibrated and periodic navigation profiles were performed. The data was stored in ASCII format file making it easy to export to CAD or GIS software.

3.2 *2008 bathymetric campaign*

One year later, another bathymetric campaign was performed, focused only in the tail of Ribarroja Reservoir (confluence Ebro-Segre Rivers). The field bathymetric campaign was carried out in December 2008, approximately a year later than the previous bathymetry. The motivation of this campaign was the study of sediment dynamics of the confluence.

This field work was carried out using the same procedure used on the previous bathymetric campaign conducted in 2007, but only a single multi-beam probe was used now. While the bathymetry of 2007 was held for the whole reservoir, in 2008 only the upstream zone was studied: set from the junction Ebro-Segre up to about five kilometers downstream. The data was gathered following the longitudinal profile (the talweg of the river) and several perpendicular cross-section profiles of the river (Fig. 4). The data was exported to ASCII file, and from it, a raster and a TIN files were obtained using ESRI ArcMap. The raster resolution was of 2×2 m cell size (Fig. 5).

3.3 *Numerical modelling*

Iber (Bladé et al. 2014) is a two-dimensional depth averaged mathematical model for the simulation of free surface flow in rivers and estuaries, developed by the Water and Environmental Engineering Group, GEAMA (University of A Coruña) and the Flumen Institute (UPC-CIMNE). Iber includes a hydrodynamic module, a turbulence module, and a sediment transport module. The application fields of the current version of Iber are: simulation of free-surface flow in rivers, flood inundation modelling, hydraulic calculation of encroachments, calculation of tidal currents in estuaries, stability of bed sediments or erosion and sedimentation due to transport of non-cohesive sediments.

The sediment transport module solves the sediment non-stationary transport equations. The equations include the bedload transport equations and the suspended sediment transport equations, coupling the bedload and the suspended load through a sedimentation-rise term.

In this work, the suspended sediment module was used. In it the convection diffusion equation of suspended sediment transport is solved together with the Shallow Water Equations and a k-ε turbulence model.

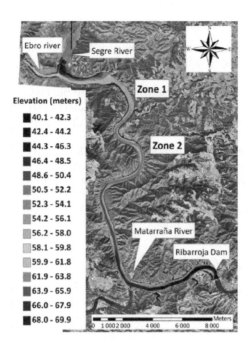

Figure 4.   Whole Ribarroja Reservoir 2007 bathymetry.

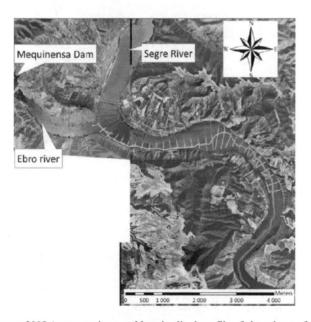

Figure 5.   Bathymetry 2008 (cross sections and longitudinal profile of the talweg of the river).

4   RESULTS

From the comparison of bathymetries, the effects of the flood could be studied and the patterns of remobilization and resuspension were obtained. Furthermore a map distribution of the sediment before and after the flood could be plotted with digital terrain models.

4.1 Comparison between 2007 and 2008 bathymetric campaigns

To analyze the obtained bathymetries the same profiles were compared. The location and the definition of these profiles, shown in Figure 6, was inherently defined by the survey profiles of 2008 bathymetry.

For the first comparison between bathymetries, a simple raster subtraction was made. In Figure 7 it can be seen that along the first few kilometers upstream there is a clear erosive trend (in black) and downstream, approximately from the 25th profile, a depositional pattern (in white) can be observed.

The proposed procedure was to subtract the earlier Bathymetry (2007) of the final Bathymetry (2008) to see the areas of erosion and sedimentation. In addition, a quantification of the magnitude of the differences was obtained. The observed differences are less than few tens of centimeters (up to 20–30 cm). Also at the cross sections it can be seen that these trends are most pronounced in the profile corresponding with the talweg (longitudinal profile) rather than the sides of the transversal profiles.

4.2 Spatial variation of sediment

In order to observe the spatial distribution of the sediment, the Ribarroja Reservoir was subdivided into segments of constant length (200 meters), and for each segment the average height was calculated (Fig. 8). The average height, sedimented (+) or eroded (−), was

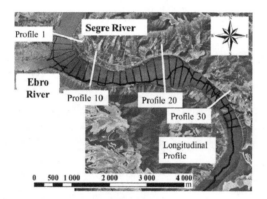

Figure 6.   Longitudinal profile and transversal cross section profiles used for the study of bathymetry 2007 versus bathymetry 2008.

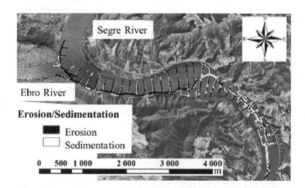

Figure 7.   Areas of erosion (black) and areas of deposition (white). The first 4000 meters downstream of the confluence Segre-Ebro have an erosive trend while downstream, the trend is mainly sedimentary.

104

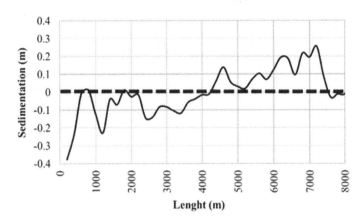

Figure 8.   Left: Average sediment accumulation for segments of 200 meters length. The areas of erosion are negative and areas of deposition are positive.

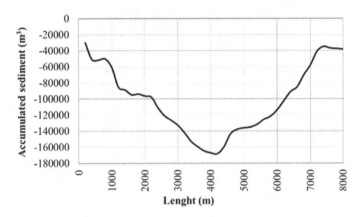

Figure 9.   Accumulated sediment curve (m³). For the first 4000 m a volume of 170,000 m³ is eroded and it is equal to the deposited volume for the next 3000 m downstream.

multiplied by the surface of each segment to obtain the average volume of erosion (−) or deposition (+) for each one (Fig. 9).

The obtained erosion volume of the first 4000 m is a relatively similar to the volume that settles in the next 3000 m downstream (Fig. 9). In other words, in this area of study a volume of about 170,000 m³ of sediment has been displaced about 4 km downstream. This is directly related to the displacement of the sediment mass front located just downstream of the confluence Ebro-Segre.

5   NUMERICAL SIMULATIONS

Iber model implements three formulations for calculating the suspended sediment concentration in equilibrium: Van Rijn (1987), Smith, J. & McLean (1977), and Ariathurai & Arulanandan (1978). The third one is the one used in this study because is the most suitable for fine cohesive sediments.

The transport module of suspended sediment was calibrated using the complete digital terrain model obtained for the year 2007, the characteristics of the sediment in the area (López et al. 2012) and the hydrograph flood that took place in 2008.

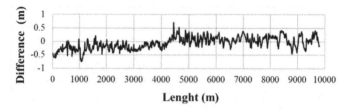

Lenght (m)

Figure 10. Longitudinal profile of bathymetry 2008 minus bathymetry 2007. The first 4000 m the values are negative and indicate erosion, whereas a positive difference indicates sedimentation from 4500 m to 10000 m.

Erosion/Sedimentation

■ Sedimentation
☐ Erosion

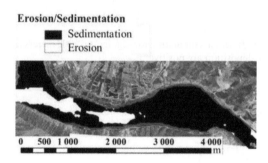

Figure 11. Areas of erosion (white) and deposition (black) calculated with Iber suspended sediment module.

After a parameter calibration process good agreement was achieved between the numerical results and to those obtained with the second bathymetry conducted in 2008 (Fig. 10). The thickness of erosion/deposition was calibrated as well as the spatial distribution of the sediment (Fig. 11).

6 CONCLUSION

By comparing the two bathymetries, the amount and the spatial distribution of accumulated/eroded sediment over one year could be assessed for the tail of Ribarroja Reservoir.

From the difference of 2007 and 2008 digital elevation models, it could be estimated that in the first 4000 m from the Ebro-Segre confluence there was a similar erosion volume than the volume that settled 4000 m downstream. In other words, in this area of study a volume of about 170,000 m^3 of sediment was displaced about 4 km downstream by the 2008 flood event.

This morphology changing process is directly related to the advancing front of the mass of sediment from Segre River entering into the Ribarroja Reservoir, which stays just downstream of the confluence (where the velocities decrease due to the Segre River mouth). Because of this reason, after the flooding event occurs the sediment were settled in the first 4500 m downstream. The mass of sediment deposited precisely in the confluence can be only removed by major flood episodes, when flow exceeds a certain velocity and it has sufficient shear stress to resuspend the sediment.

This study case was also numerically studied using the suspended sediment transport module implemented in the two-dimensional Iber modelling tool. The results after the simulations performed with Iber model were similar (in terms of depositional trends) to those obtained with the second bathymetry conducted in 2008. The thickness of erosion/deposition was calibrated for different formulations and parameters as well as for the spatial distribution of the sediment.

ACKNOWLEDGEMENTS

This paper is supported by the research FI PhD program of The Agency for Management of University and Research Grants (AGAUR) of Catalonia. The authors would like to thank ENDESA and the Water Authority of Ebro River (Confederación Hidrográfica del Ebro) for providing the daily Reservoir and Segre flow data for the study period.

Maps throughout this communication were created using ArcGIS® software by Esri. ArcGIS® and ArcMap™ are the intellectual property of Esri and are used herein under license. Copyright © Esri. All rights reserved.

REFERENCES

Ariathurai, R. & Arulanandan, K. 1978. Erosion rate of cohesive soils. Asce Journal of the Hydraulics Division, 104(HY2): 279–283.

Bladé, E. Cea, L., Corestein, G., Escolano, E., Puertas, J., Vázquez-Cendón, E., Dolz, J. & Coll, A. January–March 2014. Iber: herramienta de simulación numérica del flujo en ríos. Revista Internacional de Métodos Numéricos para Cálculo y Diseño en Ingeniería, Volume 30, Issue 1, Pages 1–10.

Garcia, M.H., 2008. Sedimentation engineering: processes, measurements, modeling and practice. *ASCE Manuals and Reports on Engineering Practice* No. 110 M.H. Garcia, ed., American Society Civil Engineering Publications.

Kawashima, S. et al., 2003. Reservoir conservation: The RESCON approach, Vol. II., Washington, D.C.

LIMNOS, 1996. Diagnóstico y gestión ambiental de embalses en el ámbito de la cuenca hidrográfica del Ebro. Embalse de Mequinenza. Report.

López, P., Dolz, J., Arbat, M., & Armengol, J., 2012. Physical and chemical characterisation of super fi cial sediment of the Ribarroja Reservoir (River Ebro, NE Spain). *Limnetica*, 31(2), 321–334.

Morris, G. & Fan Jiahua, 1998. Reservoirs sedimentation handbook, New York: McGrawHill.

Morris, G.L., Annandale, G.W. & Hotchkiss, R., 2008. Reservoir Sedimentation. In M.H. García, ed. *Sedimentation Engineering. Processes, Measurements, Modeling, and Practice. ASCE Manuals and Reports on Engineering Practice No. 110*. ASCE, p. 1132.

Palmieri, A. et al., 2003. Reservoir conservation: The RESCON approach Vol. I, Washington, D.C. Available at: http://sp.uconn.edu/~fshah/Vol1.pdf.

Prats, J. et al., 2007. A methodological approach to the reconstruction of the 149–2000 water temperature series in the Ebro River at Escatron. *Limnetica*, 26(2), pp.293–306.

Prats-Rodríguez, J. et al., 2011. Dams and Reservoirs in the Lower Ebro River and Its Effects on the River Thermal Cycle D. Barceló & M. Petrovic, eds. *The Ebro River Basin*, (July 2010), pp.77–95.

Smith, J. & McLean, S., 1977. Spatially averaged flow over a wavy surface. *Journal of Geophysial Research*, 82:1735–1746.

Strand, R.I. & Pemberton, E.L., 1982. Reservoir sedimentation. Technical Guideline for Bureau of Reclamation., Denver.

Van Rijn, L.C. 1987. Mathematical modelling of morphological processes in the case of suspended sediment transport. *Delft Hydraulics Communication* No. 382. Delft Hydraulics Laboratory. Delft, The Netherlands.

Reservoir Sedimentation – Schleiss et al. (Eds)
© *2014 Taylor & Francis Group, London, ISBN 978-1-138-02675-9*

The use of Flow Assisted Sediment Transport (FAST) to remove sediment from reservoirs in Southern California, USA

F. Weirich
IIHR Hydroscience and Engineering, University of Iowa, Iowa, USA
Department of Earth and Environmental Sciences, University of Iowa, Iowa, USA

ABSTRACT: Many of the reservoirs in Southern California are subject to relatively high sediment accumulation rates. In an effort to provide a cost effective and environmentally acceptable method of removing sediment from these reservoirs a series of operational experiments were conducted on three of the major reservoirs in this area. In two of the reservoirs an approach that might be best described as a version of hydraulic flushing, involving carefully managed Flow Assisted Sediment Transport (FAST), was employed to convey sediment out of the reservoirs. In the third reservoir a more conventional method of draining the reservoir followed by sediment removal using mechanical extraction and trucking was employed. The volume of sediment removed in these operations ranged from approximately 350,000 m^3, to in excess of 3,000,000 m^3. In total in excess of 5,000,000 m^3 were removed as part of these projects. The results of both an initial pilot study and the more extensive larger scale operational experiments, summarized in this paper, suggest the use of the carefully managed FAST approach, in appropriate situations, can be employed at a fraction of the cost of conventional removal methods such as trucking or dredging along with the added benefit of significantly lower environmental impacts.

1 INTRODUCTION

Efforts to either prevent sediment from entering or depositing in reservoirs or finding effective ways to remove sediment that has deposited in reservoirs has been a longstanding challenge. The actual scale and significance of the problem has been documented in a number of studies such as Mahmood (1987), Yoon (1992) and Shen (1999). In these studies, and others, while it is clear that the loss of annual storage capacity varies widely, with often cited values ranging from an average of 0.3% to 2.3% annually across the globe, the mostly commonly agreed upon number suggests that the loss of reservoir storage capacity on a global level averages approximately 1% per year. Moreover, it is not simply a matter of the volume of the material alone. The challenge of maintaining or restoring reservoir storage capacity and ensuring the continued operational control of outlet works has become increasingly difficult in the face of rising removal costs, increased environmental requirements, and in many cases, increasing difficulty in finding acceptable disposal sites.

The problem has been of particular concern with respect to a number of the major reservoirs in Southern California, USA, where the combination of highly erosive steep slopes, relatively steep channel gradients, sequences of often intensive storm events, a history of fire related high volume debris flows into reservoirs, and a reservoir system located immediately adjacent to a major urban area has resulted in the relatively rapid infilling of many of the major reservoirs with sediment. At the same, there are significant limitations and constraints in this area in terms of the ability to restore and maintain storage capacity associated with environmental and usage concerns and a lack of cost effective disposal sites. The scale and nature of these challenges are well documented in the reports of the Army Corps of Engineers & Los Angeles Department of Public Works, Flood Control District (ACOE/LADPW) (1994), and The Los Angeles Department of Public Works, Los Angeles Flood Control District (LADPW/LACFCD) (2013).

A wide range of efforts and methods have been developed and evaluated around the world to address these types of reservoir sedimentation issues. An overview of these efforts and methods is provided in Morris & Fan (1998) (2010) and White (2005). One approach to the challenge of reservoir sediment management which has been widely discussed involves the use of some form of flushing of sediment either from the reservoir floor (hydraulic flushing) or flushing of materials through the reservoir in such a manner as to prevent deposition (an approach classified as sediment routing by (Morris & Fan (1998) (2010) or some combination of the two approaches (which is often also described as sluicing), Aktinson (1996). Brandt (2000), Fan and Morris (1992), Shen (1999) provide helpful summaries of such efforts. Atkinson (1996), Fan & Morris (1992), Lai & Shen (1996), Shen (1999), White & Akers (2000), and Batalla & Varicat (2009) provide both an overview of the broad considerations involved in the use of flushing as a technique as well as an evaluation of the limitations and constraints associated with using this approach based on either field, laboratory, computational modeling efforts or a combination of such approaches and evaluations.

In an effort both to address some of the sediment management issues associated with the operation of several of the major reservoirs in Southern California and to evaluate the overall relative effectiveness of several different reservoir sediment management techniques, a series of operational experiments were conducted on three of the major reservoirs in the Los Angeles Flood Control District. The three reservoirs, Cogswell Reservoir, San Gabriel Reservoir and Morris Reservoir are located in the San Gabriel Mountains and are part of the drainage system of the San Gabriel River. In two of the reservoirs, the Morris and San Gabriel Reservoirs, the use of a form of hydraulic flushing involving the use of carefully managed and controlled flow assisted sediment transport to remove sediment from the reservoir floor was employed, while a third reservoir, the Cogswell Reservoir, was essentially used as "control" with sediment being removed by more conventional means. The volume of sediment removed during these operational experiments ranged from approximately 350,000 m^3 in the Morris Reservoir, which was used as a pilot study, to in excess of 3,000,000 m^3 in the other two reservoirs where full scale operational experiments were undertaken.

2   THE SETTING

The Upper San Gabriel reservoir system consists of three reservoirs on the San Gabriel River (see Fig. 1) with a combined drainage area of approx. 548 km^2. The upstream reservoir in the system, the Cogswell Reservoir, with a storage capacity of approx. 14,500,000 m^3 receives drainage from an area of 101 km^2. The San Gabriel Reservoir with a capacity of 66,500,000 m^3 receives drainage from Cogswell Reservoir and an additional drainage area of 423 km^2 for a total of 524 km^2, while the Morris Reservoir with a storage capacity of 40,100,000 m^3 receives drainage both from the Cogswell and San Gabriel Reservoirs plus an additional drainage area of 23 km^2. Despite the relatively limited size of the drainage area of these three reservoirs the latest estimates by the LACDPW/LACFCD (2013) indicate that this system delivers approximately 40% of all the sediment deposited in the entire LACDPW/LACFCD reservoir system of some 14 major reservoirs. This is the result of a combination of a relatively high series of mountains (exceeding 2500 m in height) steep drainages with slopes averaging above 55% in the majority of the Upper San Gabriel drainages, steep gradient streams with little intermediate storage capacity, relatively highly erodible soils associated with rapid uplift of the San Gabriel Mountains, and a Mediterranean climate pattern of often intense winter storm sequences often accompanied by high intensity rainfall events, Weirich (1994a).

The operational plan for this reservoir system (LACDPW/LACFCD, 2013) seeks to maintain an operational storage capacity of approx. 61,675,000 m^3 not including the capacity of the Morris Reservoir. At the time of the operational experiments discussed in this paper

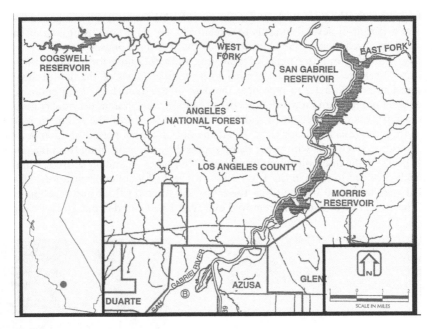

Figure 1. Location of the three Upper San Gabriel River reservoirs (adapted from COE/LACDPW, 1994) (Weirich, 1994a).

Cogswell Reservoir held approximately 4,300,000 m³ of sediment. San Gabriel Reservoir contained 11,700,000 m³ of sediment and Morris Reservoir held 8,700,000 m³ of sediment. The resulting reduction in storage capacity of these reservoirs to approx. 70%, 82% and 78% of their design capacity prompted a concerted effort to both restore more of the operational storage capacity and also evaluate the feasibility and cost effectiveness of employing different approaches to the sediment management issue. The Morris and San Gabriel Reservoir operational experiments using carefully managed FAST and the Cogswell mechanical cleanout operation were part of that effort.

3   THE MORRIS DAM PILOT STUDY

In the initial phase of this series of sediment removal efforts a pilot study was undertaken on the Morris Reservoir. The goals of the pilot study was to assess the overall cost effectiveness of removing a significant amount of sediment (approximately 350,000 m³) from the reservoir using a version of hydraulic sluicing involving the use of carefully managed FAST. The plan was to use carefully controlled flows from the San Gabriel Reservoir to remove sediment from the Morris Reservoir floor and then make use of the same flows to convey the sediment down the San Gabriel River to disposal sites located out in the San Gabriel Valley. As part of this effort a careful monitoring program of the process was undertaken.

   The sequence of events associated with this effort involved: a) removal of at-risk species from a 10 km downstream portion of the river below the dam as well as the adjacent riparian corridor; b) removal of fish from the Reservoir itself prior to draining of the reservoir; c) the draining of the reservoir; d) preparation of the outlet works for sediment removal process; e) installation of monitoring equipment to track sediment concentration levels during the experiment and volume removal estimates and a GIS based assessment of reservoir floor conditions pre- and post the experiment as well as sediment deposition patterns in the downstream areas; f) post experiment downstream channel and habitat restoration; and g) the assessment of the fate of sediment conveyed down the river system.

In preparation for the pilot study in the summer of 1991 the reservoir was gradually drawn down over a period of several months to a relatively low pool level using one of the main valves. The actual sediment removal process took place in three phases.

Initially, beginning in late August of that year and continuing for a period of approximately three weeks residual flow associated with the final stages of the dewatering process was channeled through a 1.2 m pipe near the base of the dam (Fig. 2a–b) and a sediment load sampling program and a continuous density monitoring system previously installed on the 1.2 m pipe was employed to provide a continuous record of the sediment load being removed from the reservoir floor (Fig. 3). At the outset of this period of final dewatering, flow rates exceeded 8.6 m^3/s with water being supplied by water draining from the remaining standing pools on the reservoir floor, groundwater and other secondary sources. But the flow rate quickly dropped to rates in the 1.0–1.3 m^3/s range. Sediment concentrations of 10–12% by weight were measured for the flows traversing the reservoir floor and reaching the pipe intake. The initial rate of sediment removed exceeded 15,300 m^3/day but quickly dropped to

 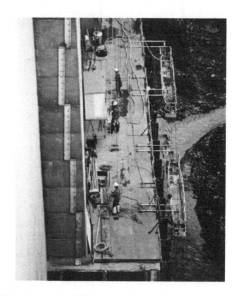

Figure 2.   (a) Reservoir at end of dewatering. The FAST intake area is visible at dam base. (b) Temporary platform above intake pipe at dam base with flow coming from upper right.

Figure 3.   The 1.2 m exit pipe at base of dam carrying FAST flow. Sensors to continuously measure sediment concentration were mounted in this pipe.

a 2,300–4,600 m³/day rate. Over this three week period of completing the dewatering process, approximately 63,500 m³ of sediment was removed from the reservoir floor. During the first days of this process most of the material being moved was in the silt/clay range but the amount of sand and larger fractions gradually increased as headward erosion by the stream channel forming on the reservoir floor and draining the reservoir provided larger quantities of materials from further up the reservoir floor. The vast majority of this material was deposited in the plunge pool of the dam and other pools not far downstream of the dam.

The second phase involved the actual operational experiment and entailed a carefully managed release of water from the upstream San Gabriel Dam to provide the (FAST) to remove sediment from the reservoir floor and transport it through the 1.2 m pipe and downstream of the dam. During this phase of the experiment, which lasted for a period of approximately 11 days, some 270,000 m³ of material was removed from the reservoir floor with an average removal rate of 20,600 m³/day and a peak value in excess of 24,500 m³/day, with sediment concentrations in the flows exceeding 200 G/L at times (Fig. 4). The sampling of sediment indicated that a much larger portion of the material being moved was in the sand/gravel/pebble range than had been moved in the three week final drawdown operation. A much larger amount of the material moved during this period was deposited beyond the plunge pools in a portion of the channel extending some 10 km downstream of the dam.

The final phase of the sediment removal effort covered a period of approx. four more weeks after the FAST portion came to close with the end of the controlled release of sediment entraining flows from the San Gabriel Dam. During this final phase residual reservoir surface drainage, groundwater and minor leakage from the San Gabriel Dam provide some flow that continued to remove sediment from the reservoir floor. The daily rate of removal was in the 750 m³/day range with a total amount of sediment being removed of approx. 19,000 m³.

Careful mapping and monitoring of the downstream fate of the sediment removed from the Morris dam during this experiment indicated that of the approx. 350,000 m³ of material removed during the overall experiment approx. 59,000 m³ was carried more than 10 km downstream during the experiment to engineered structures further down the river with

Figure 4. A portion of the floor of the Morris Reservoir at the end of the sediment removal experiment showing the lateral extent and depth (approx. 10 m) to which material was removed.

the remainder being deposited in the pools and channel areas immediately downstream of the dam. Subsequently, and as part of the overall project, a series of controlled releases of 15 m³/s to 45 m³/s were made in mid-January over the course of several days involving the release of approx. 6,900,000 m³ of water. These releases, along with more conventional releases, were undertaken with the intent of re-entraining the material deposited in the channel downstream during the experiment in order to carry that material out of the natural channel portion of the river downstream of the dam and convey it further downstream into the fully engineered portion of the San Gabriel River in the interest of habitat restoration. Monitoring of this effort indicated that a combination of the downstream flushing operation and normal dam releases over a period of months left approx. 53,500 m³ of the original 350,000 m³ in the natural 10 km portion of the channel downstream of the dam, the vast majority of those 53,500 m³, consisting largely of larger fractions of coarse sand up to cobbles and small boulders, were located in the two major plunge pools immediately downstream of the dam. Total cost of the sediment removal effort, including habitat restoration, were estimated at approx. $1.30/m³ (less than 1/5 the approx. cost of equivalent mechanical removal which at the time ranged up to above $30/m³).

4   THE SAN GABRIEL RESERVOIR OPERATIONAL EXPERIMENT

Based on the success of the pilot study a second, larger scale operational experiment was carried out on the San Gabriel Reservoir following the same approach of employing carefully managed FAST to remove sediment accumulated on the reservoir floor. In this instance, following a dewatering process, carefully controlled releases from the Cogswell reservoir, upstream of the San Gabriel Reservoir provided the sediment entraining flow. The sediment laden flows were then routed to a low level 1.8 m sluice tunnel located at the base of the dam that had been incorporated in the original design of the dam and then into the Morris Reservoir immediately downstream of the San Gabriel Reservoir via a tributary canyon which received flow from the exit of the sluice tunnel and then drained directly into the Morris Reservoir.

The outflow from the sluice tunnel was continuously monitored throughout the operation using an integrated system incorporating sediment concentrations sensors/a discharge

Figure 5.   Floor of the San Gabriel Reservoir floor early on in the operation.

114

Figure 6.   Floor of the San Gabriel Reservoir floor later in the process. The entrance to the old sluice tunnel used in this operation is located in the upper right corner at the base of the dam.

Figure 7.   The use of equipment to assist in increasing the sediment concentration effectively doubled the sediment concentration of the material being removed from the floor area.

monitoring system and coordinated water and floor sediment sampling throughout the duration of the operation which extended for a period of some 5 months from early August to late December, 1992.

The monitoring effort indicated that a total of 1,691,000 m³ of sediment were removed during this operation. This value was confirmed by a separate and blind computation of the

Figure 8. The exit area of the 1.8 m diameter tunnel used to convey the removed material. Instrumentation to continuously monitor and sample the sediment load being carried was located here.

actual removal volume using a GIS based evaluation of pre-operation and post operation photogrammetric surveys of the change in the reservoir floor. The independent GIS based volumetric assessment of the amount of sediment removed indicated that 1,737,000 m³ had been removed during this operation. The relatively close agreement (within 2.5%) confirmed the reliability of the removal monitoring techniques.

Continuous monitoring of sediment concentrations during the operation indicate that sustained concentrations of 5%–6% by volume of sediment were generally maintained. These levels rose to 10%–12% by volume when coordinated mechanical assistance was used to boost the entrainment and transport rate resulting in sustained maximum sediment levels of above 264 G/L. These values translated into an average sediment removal rate of 11,850 m³/day or the life of the project with a range from 2,300 m³/day to in excess of 40,500 m³/day. The wide range was the result of a combination of factors: a) variation in flow levels released from Cogswell Dam (ranging from 1 m³/s −5.7 m³/s); b) the use and amount of mechanical flow assisting equipment used at different times; c) enhance discharge from rainfall events; d) source; e) location and f) the nature of San Gabriel Reservoir flow materials being entrained and transported.

Efforts to control both the levels of the sediment concentrations and the selective removal of materials from specific locations, such as from the area surrounding the intake works were also successful. The cost of the operation was on the order of $0.93/m³.

5   THE MECHANICAL EXCAVATION OF THE COGSWELL RESERVOIR

In parallel with the San Gabriel FAST operational experiment, the third reservoir, Cogswell, also underwent a sediment removal effort. But in the case of Cogswell, more conventional mechanical removal techniques were employed. In part, this was undertaken to allow for a comparison with the FAST approach that had been used in the other two reservoir. This approach was also used in order to minimize downstream impacts below the dam because of the sensitive nature of the habit of the downstream channel and riparian corridor between

116

the Cogswell and San Gabriel Reservoirs. The Cogswell mechanical excavation spanned a period of over four years from 1991–1996 and began with a careful drawdown procedure and a channel monitoring procedure. In the fall of 1992 approx. 428,000 m³ and in the 1994–1996 period an additional 2,689,000 m³ were removed by excavation and trucked to a nearby disposal site. All told approx. 3,102,000 m³ were removed over this period using mechanical means. The overall cost for this effort was estimated to be approximately $6.50 m³.

6 THE COMPARISON OF METHODS

While all three of the projects described above were successful in terms of removing sediment and restoring reservoir capacity, the use of a carefully managed FAST approach in both Morris and San Gabriel Reservoirs were conducted in a shorter period of time at much lower overall cost per m³ and somewhat lower environmental impacts (such as increased air pollution) than the Cogswell mechanical excavation. This is not to say the FAST approach is not without limitations and constraints. The downstream impacts inherent in this approach must be carefully considered, especially with respect to increasing concern regarding environmental impacts and habitat restoration requirements that may need to be incorporated in the overall costs of using this approach. There are also clear limits on the size fractions of the material that can be effectively removed with this approach, a constraint not generally an issue for some other techniques such as mechanical excavation. The amount of water required for the FAST approach both for direct entrainment and transport as well as possible post operation flushing of sediment deposited in downstream reaches of a dam in order to restore downstream habitat must also be considered. In some situations, as was the case in both of the FAST operations described in this paper, provision can be made to recover the water used in facilities further downstream.

7 SUBSEQUENT SEDIMENT MANAGEMENT EFFORTS

Based on the results of the initial FAST pilot study on the Morris Reservoir and the subsequent larger FAST operational experiment on the San Gabriel Reservoir in 1998 a second FAST type operation was conducted resulting in the removal of 1,604,000 m³. This was followed in March of 1999 with a sediment flushing operation, similar to that conducted in the pilot study to clear the channel downstream of Morris Dam of sediment accumulated in the pools and floodplain zone (LACDPW/LACFCD, 2013). An environmental monitoring study of the post operation channel recovery effort (LACDPW/LACFCD, 2013) indicated that within 2–3 years of the operation the downstream habitat had been restored to pre-FAST conditions. Going forward (LACPW/LACFCD, 2013) planning for sediment removal involves a mix of strategies for sediment management in which FAST or variations on the approach will likely be one of the options available.

8 CONCLUSIONS

The results of both an initial pilot study and the more extensive larger scale operational experiments, summarized in this paper, suggest the use of the carefully managed FAST approach, in appropriate situations, can be employed at fraction of the cost of conventional removal methods such as trucking or dredging along with significantly lower environmental impacts.

REFERENCES

Atkinson, E. 1996. The feasibility of flushing sediment from reservoirs. H R Wallingford, U.K. Report No. OD 137, pp. 21.

Batalla, R.J. & Variant, D. 2009. Hydrological and sediment transport dynamics of flushing flows, implications for management in large Mediterranean Rivers. *River Research and Applications,* 25: 297–314.

Brandt, S.A. 2000. A review of reservoir desiltation. *International Journal of Sediment Research,* 15(3): 321–342.

Fan, J. & Morris, G.L. 1992. Reservoir sedimentation. II. Desiltation and long-term storage capacity. Journal of Hydraulic Engineering, 118(3): 370–384.

Los Angeles County Department of Public Works, The Los Angeles Flood Control District. 2013. Sediment Management Strategic Plan 2012–2032. Los Angeles, CA.

Lai, J. & Shen, H.W. 1996. Flushing sediment through reservoirs. *J. of Hydraulic Res.* 34(2): 237–255.

Mahmood, K. 1987. Reservoir Sedimentation: Impact, extent, and Mitigation. World Bank Technical Paper No. 71. The International Bank for Reconstruction and Development. Morris, G.L. & Fan, J. 1998. *Reservoir Sedimentation Handbook*. New York: McGraw-Hill.

Morris, G.L. & Fan, J. 2010. *Reservoir Sedimentation Handbook: Design and Management of Dams Reservoirs, and Watershed for Sustainable use (electronic version)*. New York: McGraw-Hill.

Shen, H.W. 1999. Flushing of sediment through reservoirs. *J. of Hydraulic Res.* 37(6): 743–757.

US Army Corps of Engineers & Los Angeles County Department of Public Works. 1994. San Gabriel Canyon Sediment Management Plan, Los Angeles, CA.

Weirich, F. (1994a). Morris Reservoir Pilot Sluicing Project Final Report on the Sediment Monitoring Study. (unpublished report).

Weirich, F. (1994b). San Gabriel Dam—Emergency Sediment Removal Monitoring Program Final Report. (unpublished report).

White, W.R. & Ackers, J. 2000. Guidelines for the Flushing of Sediment from Reservoirs. Report SR 566, H R Wallingford, U.K.

White, W.R. 2005. A review of Current Knowledge World Wide Water Storage in Man-Made Reservoirs. FR/R0012, Foundation for Water Research Allen House, Liston Road, Marlow. pp 40.

Yoon, Y.N. 1992. The state and perspective of the direct sediment removal methods from Reservoirs. International Journal of Sediment Research, 7(20): 99–115.

Reservoir Sedimentation – Schleiss et al. (Eds)
© *2014 Taylor & Francis Group, London, ISBN 978-1-138-02675-9*

Two approaches to forecasting of sedimentation in the Stare Miasto reservoir, Poland

T. Dysarz
Department of Hydraulic and Sanitary Engineering, Poznan University of Sciences, Poland
Institute of Meteorology and Water Management, Poland

J. Wicher-Dysarz
Department of Hydraulic and Sanitary Engineering, Poznan University of Sciences, Poland

M. Sojka
Institute of Reclamation, Land Improvement and Geodesy, Poznan University of Sciences, Poland

ABSTRACT: The main idea of the presented paper is a comparison of the two approaches to modeling reservoir sedimentation. The analysis is preformed on the basis of the actual data and simulations performed for the real lowland reservoir, Stare Miasto, located in central part of Poland. Both compared methods are one dimensional. The first is the basic approach based on the sediment routing model described by several transport formulae. The second is a newer approach called the Sediment Impact Analysis Method (SIAM). The purpose of the analyses presented is evaluation of usefulness of the methods for design of a lowland reservoir. Such elements as (1) sediment distribution along the reservoir, (2) need for geometry update in SIAM, (3) number of computations needed to obtain valuable results, are taken into account. The robustness of obtained results is verified on the basis of comparisons with field measurements. The general conclusions state that the new approach SIAM is very interesting and characterized by huge computational efficiency in comparison with the standard sediment routing model. However, the SIAM method is still not able to provide as reasonable results as the standard one. Hence, further developments in this area are necessary.

1 INTRODUCTION

The aim of the presented research is to compare the two methods for forecasting of sedimentation in a lowland reservoir. The chosen object is the Stare Miasto reservoir on the Powa river located in central part of Poland. Two 1D methods are used for modeling of the sedimentation process. The first is the standard sediment routing model. The second is the Sediment Impact Analysis Method (SIAM). The implementation of the methods and the obtained results are compared and discussed. The analysis is made taking into account the usefulness of the methods for design of lowland reservoirs.

One of the main problems related to the assessment of reservoir sedimentation is effective simulation of this process. The most popular of 1D/2D models describing sediment transport and deposition are based on the same construction as flood wave propagation models (e.g. Papanicolaou et al., 2008). The fundamental conservation laws are integrated in a relatively small control volume under several assumption. The most important are short time intervals and continuity of all parameters in the resulting partial differential equations (e.g. Wu, 2008). Such models work well for relatively short time flood phenomena of duration from one week to two months at most. The duration of reservoir sedimentation phenomena is much longer, from a few to fifty years. The time steps, which guarantee the stability of computations are too small to perform such simulations effectively. The problem of different time scales was noticed and described in the papers written by Cao et al. (2007) and Cao et al. (2011). The additional side of the problem is huge uncertainty of parameters describing the features of the sediment transport model (e.g. Bogardi et al., 1977; Salas & Shin, 1999). Hence, such models may be properly calibrated and validated only when confronted with the existing reservoirs. During the design process, the forecast of sedimentation over long period requires repetition of simulations taking into account variability of inflows as well as uncertainty of parameters.

The problems with effectiveness of sediment transport simulations lead to simplified approaches. There are numerous examples of them in the scientific literature varying from more physically based models, e.g. Rahmanian & Banihashemi (2011), to those based on black-box idea, e.g. Yitian & Gu (2003), Nourani (2009). One of the most interesting approaches seems to be the so-called Sediment Impact Analysis Method (SIAM). The method was developed by Mooney (2006). The primary idea of this approach is the sediment budget tool, balancing the sediments inflows from watershed to the river. The computations of sediment aggradation and degradation are made along one dimensional river reach. Although, there is no update of geometry related to sediment transport results, the construction of the method seems to be applicable to the problem of reservoir sedimentation.

The idea of this paper is to compare the performance of the standard sediment routing model with that of the simplified SIAM one. Several scenarios of the sediment routing model are processed and then compared with the results of SIAM. The assessment is made taking into account the sediment distribution along the reservoir, need for geometry update in SIAM and also the amount of computations needed to obtain valuable results. Finally, the results are compared with direct measurements of geometry in the chosen reservoir.

The paper consists of 5 sections. The first is the introduction. Then the materials used in our investigation are described. In the third section the methods applied are explained. The results are discussed in the fourth section. The conclusions are presented in the last section.

2 MATERIALS

The Stare Miasto reservoir is located on the Powa river in the central part of Poland. The main dam is located to the south of Konin city. The reservoir is a relatively new object, built in 2006. Its length is 4.5 km and the area of inundation in normal conditions is 90.68 ha. The total capacity of the reservoir is 2.159×10^6 m^3, but the capacity used for water supply is 1.216×10^6 m^3. Highway A2 is narrowing the active flow cross-section in the central part of the reservoir. The dam splitting the object into the main and upper part is located upstream of the bridge (Fig. 1). The upper dam includes a small sluice. The area of the upper part is 27 ha. The capacity of this part is 0.294×10^6 m^3 (Woliński & Zgrabczyński, 2008). The depth of the reservoir varies from 1.2 m in the upper part to 5.7 m near the main dam.

The upper part of the reservoir plays a specific role. It is used to collect sediment and protect water quality in the main part from degradation. It is expected that the sediment transported with the inflowing water is settled in the upper part of the reservoir. After some time the upper part should develop conditions good for vegetation growth. This enables the removal of pollutants from water or their deposition with the sediments.

The Stare Miasto reservoir is multi-purpose and works in the annual cycle. The main part of the reservoir is used in ordinary way. It includes water supply capacity, the dead zone as well as the flood protection capacity and the hydraulic flood protection zone. The water stored is used mainly for irrigation and protection of biological life in the Powa river. An important purpose is the flood protection of Konin city. The reservoir is additionally used for tourism and fishery.

The basic data used for presented analyses are the geometry of the reservoir, water surface levels measured at the main dam as well as the inflows to the reservoir. The preparatory data include analysis of measurements and topographic maps in the scale 1:2000 from 2006 (Woliński & Zgrabczyński, 2008). On the basis of the historical maps, the digital terrain model for the reservoir was prepared. Then the DTM was processed using ArcGIS 9.2 with HEC-GeoRAS set of

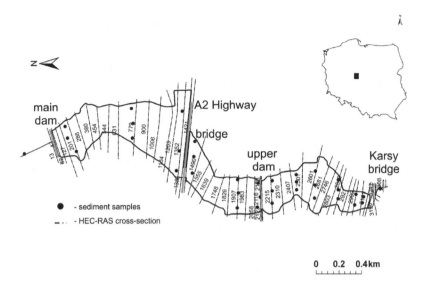

Figure 1. The Stare Miasto reservoir.

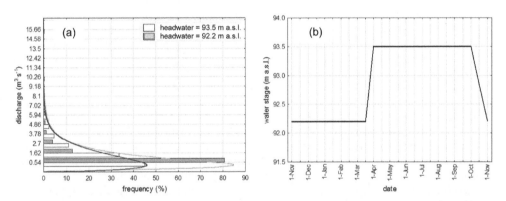

Figure 2. Hydrology and water management: (a) Relative frequency curve of discharges observed at the Posoka gauge station in the Powa river; (b) Typical changes in water levels in the Stare Miasto reservoir.

tools. It enabled definition of basic geometry for hydraulic and sediment computations. Direct measurements were also used for determination of current reservoir bathymetry. Such a survey was made in September 2013. The field survey was made using Echotrac CVM, which is an echo sounder produced by Teledyne Technologies Company.

The average annual unit outflow from the watershed in the dam cross-section is estimated as 3.52 dm³/(km² s). The closest gauge station is Posoka controlled by the Institute of Meteorology and Water Management (IMGW). The data collected at this gauge station for the period 1975–2009 are available. The inflow varies from 0.012 to 42.6 m³/s. Two discharge frequency curves were prepared on the basis of these data. The first represents variability of discharge in the months with minimum headwater in the dam. The second is prepared for the seasons with normal water level in the reservoir. The basic frequencies are shown as bar graphs in Figure 2.a. The water surface level varies from the minimum elevation of 92.70 m a.s.l. to 94.00 m a.s.l. Normal water level is 93.50 m a.s.l. (Woliński & Zgrabczyński, 2008). The lower levels are kept during the first part of the year (Fig. 2.b). From April to October the reservoir is used with normal water level.

The data set is completed with analysis of 36 bed sediment samples. They were collected at the sites marked in Figure 1 (center). The sieve analysis is used to determine the mass ratio in classes of diameters compatible with Polish regulations. The bed samples are used to make the standard gradation curves. The particles coarser than 2 mm were not observed in the samples. In general the sediments analyzed were classified as fine or very fine sands. These data were previously used and described, e.g. Dysarz & Wicher-Dysarz (2013a), Dysarz et al. (2013b).

3 METHODS

Two methods were used for estimation of sedimentation in the Stare Miasto reservoir. These were 1D methods namely the standard sediment routing and the Sediment Impact Analysis Method (SIAM). The first is based on the well known Exner's equation (e.g. Exner, 1920, 1925; Parker, 2004; Brunner, 2010). The second is a modern computation technique invented by Mooney (2006). Both of them are implemented in HEC-RAS package for hydraulic computations. The newest version of this software was used for computational tests, namely HEC-RAS 5.0 Beta (Brunner, 2010). The methods are described briefly below.

The HEC-RAS is the well known 1D hydraulic software including flow modules as well as sediment transport and water quality algorithms. The newest version 5.0 Beta is used in this research. The first step in configuration of any HEC-RAS computations is preparation of the modeled system geometry. The cross-sections and their layout is prepared by means of HEC-GeoRAS (Fig. 1). The river stations are at the same distance from the reservoir outlet to the particular cross-section. In general there are 38 cross-sections inside the reservoir with river stations varying from 47 to 3100. The sediment transport module consist of two elements: (1) quasi-unsteady flow model, (2) sediment transport algorithm. In the quasi-unsteady flow module the boundary conditions and water temperatures are set. Two types of simulations were made, namely one year and eight year computations. In the one year simulation the inlet boundary condition is a single annual hydrograph of flows. In the outlet, the stage hydrograph representing water heads in the reservoir is imposed. Thirty five simulations were performed, each one represented one hydrological year. Also ten eight-year simulations were carried out. In these simulations, individual hydrological years were randomly chosen to compose the inlet boundary conditions. The downstream boundary was composed by simple repetition of previous stage hydrographs. The sediment transport algorithm was configured by definition of soil samples, their assignment to cross-sections, definition of admissible erosion depths and sediment boundary conditions. The last element was set as equilibrium load in all simulations.

The second approach used in this study is SIAM. This is one dimensional sediment budget tool, which differs significantly from the typical sediment routing model. The method was developed as part of the Mississippi Delta Headwaters project as a joint effort of the Engineering Research and Development Center (ERDC) and Colorado State University (Little & Jonas, 2010). The first implementation of SIAM was made by Mooney (2006). SIAM is also available in the versions of HEC-RAS 4.1 and next (Little & Jonas, 2010; Brunner, 2010). In this study two SIAM models were configured. Both of them were used to calculate aggradation and degradation inside the reservoir. Hence, the sediment reaches were defined between river stations from 47 to 3100. There were two configurations of sediment reaches used for simulations. In the first there were 38 sediment reaches. Each of them was defined between two subsequent cross-sections. The second SIAM configuration was simplified and included only 14 sediment reaches representing consistent areas of the reservoir, e.g. inlet zone, upstream of a structure, downstream of a structure, etc. The averaged soil samples were assigned to each reach in both configurations. The important element of SIAM implementation is discharge frequency curve (Fig. 2.a). The curves were made for the two states of the reservoir (1) headwater of 92.2 m a.s.l., (2) or 93.5 m a.s.l.. These water levels and their duration was consistent with the rules of water management in the Stare Miasto reservoir (Woliński & Zgrabczyński, 2008).

Both algorithms, sediment routing as well as SIAM, were used with the same sediment transport functions, (1) Engelund-Hansen, (2) Meyer-Peter & Müller (MPM). The first of them was designed for sand transport, which corresponds to the actual conditions in the reservoir studied. The second one was a more general formula used mainly for bed load transport. It was used for comparisons (e.g. Yang, 1996). Other transport functions available in HEC-RAS were notused, because they are not suitable for fine sediments, e.g. Ackers-White, or they have occurred unstable, e.g. Yang. The results of sediment mass transported calculated by sediment routing module were also used as the sources of sediment for the most upstream reach for SIAM computations.

The results obtained are presented as distribution of invert changes along the reservoir profile. The term "invert" is used in a way consistent with the manuals of the HEC-RAS (e.g. Brunner, 2010). It simply means the minimum bottom elevation. The values of invert changes are obtained directly from sediment routing procedure. The standard SIAM results include aggradation or degradation of sediments in a year for particular reaches. These values are recalculated into invert changes assuming that deposits/removals are uniformly distributed along the cross-section bed. This assumption permits definition of the area of deposition/erosion. Than the invert change may be calculated on the basis of bulk density. The contraction related to fine sediments was also taken into account.

4   RESULTS

The results are presented in Figures 3–6 and some raw values are shown in Table 1. The presentation of results in all figures is split into two parts. In the left figure, the results for the inlet part of the reservoir are presented. The rest of results are presented in the right figure. This division had to be made because the values shown in the two parts differ significantly.

The first graphs (Fig. 3) present average invert changes obtained from the sediment routing simulations for one year time horizon. The results are expressed in centimeters, while the river stations are presented in meters. The sediment transport algorithms based on the

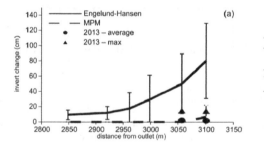
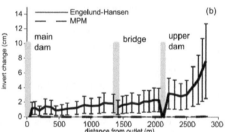

Figure 3.   Average invert changes simulated by sediment routing algorithm with Engelund-Hansen and MPM formulae: (a) inlet part—river stations 2849–3100; (b) river stations 47–2849.

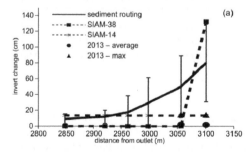
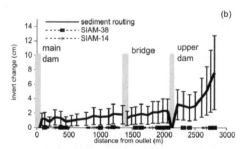

Figure 4.   Comparison of invert changes for sediment routing and SIAM with Engelund-Hansen formula: (a) inlet part—river stations 2849–3100; (b) river stations 47–2849.

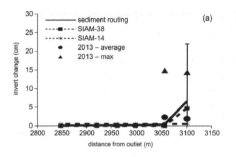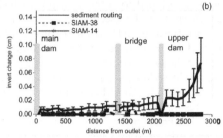

Figure 5. Comparison of invert changes for sediment routing and SIAM with MPM formula: (a) inlet part—river stations 2849–3100; (b) river stations 47–2849.

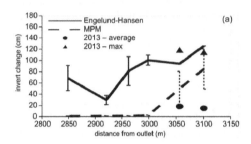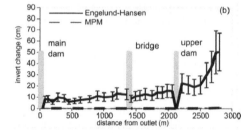

Figure 6. Comparison of invert changes between measurements and long period simulation of sediment routing: (a) inlet part—river stations 2849–3100; (b) river stations 47–2849.

Table 1. Values of invert changes for selected cross-sections.

| | Sediment routing | | | | SIAM-38 | | SIAM-14 | |
| | Engelund-Hansen | | MPM | | Engelund-Hansen | MPM | Engelund-Hansen | MPM |
RS (m)	Invert change (cm)	Standard deviation	Invert change	Standard deviation	Invert change			
3100.57	80.24	49.46	6.64	15.31	132.53	4.70	13.39	0.47
3056.61	50.19	39.21	0.39	0.44	0.01	0.00	13.39	0.47
2848.61	9.21	6.19	0.09	0.05	0.00	0.00	13.39	0.47
2501.23	3.01	1.97	0.02	0.01	0.00	0.00	0.00	0.00
2122.78	Upper dam							
1983.07	2.34	1.64	0.02	0.01	0.00	0.00	0.00	0.00
1747.59	1.72	1.25	0.01	0.01	0.00	−0.01	0.00	0.00
1400.51	Bridge							
1351.90	1.80	1.26	0.01	0.01	−0.03	0.00	−0.01	0.00
899.65	1.61	1.16	0.01	0.01	0.00	0.00	0.01	0.00
454.36	0.85	0.64	0.00	0.00	0.00	0.00	0.00	0.00
25.29	Main dam							

Engelund-Hansen and MPM formulae are compared. The results of the first formula is denoted as continuous line. The results of the second is represented by dashed line. In both cases the standard deviations from averages are marked as vertical error bars. The results of measurements are denoted as large black dots and triangles. Such results are provided only for two river stations in the inlet, namely 3100 and 3056. The spatially distributed results of

124

measurements are presented as average and maximum invert changes in each cross-section. The structures such as upper dam, the bridge and the main dam, are marked by vertical gray lines.

The sediment routing with the Engelund-Hansen formula gives greater results than the same algorithm with the MPM formula. The results of measurements are closer to the latter. This observation is in contrast to expectations as the Engelund-Hansen formulae was derived for finer sediments and expected to provide better results. The only explanation is good theoretical and experimental basis of MPM and complex nature of sediment transport in lowland rivers and reservoir. Fine sediments transported as suspended load in a mountain river become bed load in lowland streams.

The results presented in Figure 4 are comparison of sediment routing and two SIAM models for cases with Engelund-Hansen formulae. The invert changes obtained from sediment routing are displayed as a continuous line with error bars as shown in Figure 3. The SIAM results are dashed lines with points marked as rectangles (SIAM-38) and stars (SIAM-14). The graph prepared for the inlet part (left) includes also results of measurements. The right graph presents also structures. In both cases the graphical elements are the same as those used in Figure 3.

It is well seen that invert changes in the first cross-section of the reservoir resulting from SIAM-38 are greater than those obtained from sediment routing. In the subsequent cross-sections the amount of sediment deposited is smaller and invert changes are also not so great. The results of SIAM-14 show a more uniform distribution of sediments in the inlet part of the reservoir.

Similar results are presented in Figure 5, drawn for the Meyer-Peter & Müller (MPM) equation sediment transport formulae used. The vertical scales of the graphs are different, because the invert changes obtained in these cases are smaller. When this formula is used, the SIAM-38 results are more similar to those provided by the sediment routing algorithm.

Because the values presented in Figures 3–5 significantly differ in scale, some of them may be not visible well. Hence, the results for selected cross-sections are presented in Table 1. The first column consists of river stations numbers of the selected cross-. The table is composed in such a way that a comparison of sediment routing and SIAM models is clearly visible. The invert changes are expressed in centimeters. The results presented in Table 1 confirm those shown in the graphs.

The average invert changes obtained from long period simulations of sediment routing in the reservoir are shown in Figure 6. The denotations and their meaning are the same as those in Figure 3. The obtained invert changes are compared with those resulting from field measurements made in 2013. Once again the sediment routing with MPM formulae gives results closer to measurements, though, the difference between Engelund-Hansen and MPM is not so great. However, the relative compatibility between measurements and long term computations should suggest that all sediment routing simulations are reliable.

In general all SIAM results suggest huge accumulation in the inlet cross-section of the reservoir. The distribution of sediments is better, when sediment reaches are longer. Such effects are caused by the lack of any sediment redistribution or update of hydraulic conditions. The sediment redistribution is better, if the standard sediment routing algorithm is used.

5 CONCLUSIONS

In the paper a comparison of two 1D methods for assessment of reservoir sedimentation is presented. The first is the standard sediment routing model used in multi-scenario scheme. The second is the adapted SIAM algorithm. The main ideas of the methods tested are presented in section 1 and 3 of the paper. The configuration of the methods is briefly described in section 3. In both algorithms two sediment transport functions are used. These are (1) Engelund-Hansen and (2) Meyer-Peter & Müller (MPM). The choice of such functions is explained in section 3. The methods are tested on the basis of data collected and measurements from the Stare Miasto reservoir. The object as well as the data used are presented in section 2. The main element of comparisons are invert changes. The results are presented and briefly discussed in section 4. In the same section small validation of results on the basis of field measurements and long term simulations is shown.

The results presented indicate significant differences between two methods used. The differences are also visible between the transport functions applied independently of the algorithm tested. In general, the sediment routing with the Engelund-Hansen function shows greater deposition than the results obtained with the MPM formula (Fig. 3). This tendency is seen along the whole reservoir. The long period simulations show the MPM results as closer to field measurements (Figs. 3 and 6). In general, the SIAM approach shows greater irregularity in sediment deposition. The main accumulation is indicated in the inlet part of the reservoir (Figs. 4 and 5). When the Engelund-Hansen formula is applied the invert change resulting from the SIAM algorithm with 38 sediment reaches are much greater than those calculated by the sediment routing. The increase in the sediment reaches lengths and decrease in their number causes more regular distribution of sediments at least in the inlet part of the reservoir. However, the results of SIAM-14 still differ much from the sediment routing results (Figs. 4 and 5).

There are some pros and cons of the two methods applied. The sediment routing computations require complex configuration. It is time consuming, but provides more reasonable results. The results are closer to field measurements. The differences may be explained by application of 1D simplification of flow and sediment transport phenomena. On the other hand, the SIAM algorithm is simpler for configuration. The computations are faster and the results do not need to be repeated several times. However, the simplifications introduced to construct the method cause some non-physical effects. Significant irregularity in the sediment accumulation is caused by the arrangement of calculations in SIAM. The direct reason is the lack of any sediment redistribution between the sediment reaches. Such a role is played by the update of geometry and hydraulics in the sediment routing algorithm.

Although, the idea of SIAM seems to be promising, this method still needs some improvement. On the other hand, application of the sediment routing requires too much time consuming computations to be effectively used in ordinary problems such as reservoir design and prediction of capacity changes due to sedimentation. Further research in this area is necessary.

ACKNOWLEDGEMENT

The research was supported by National Science Centre in Poland as a part of scientific project "Initial sedimentation part in small lowland reservoirs: modeling and analysis of functionality", contract no. N N305 296740.

REFERENCES

Bogardi I., Duckstein L., Szidarovszky F., 1977. Reservoir sedimentation under uncertainty: analytic approach versus simulation., Hydrological Sciences-Bulletin-des Sciences Hydrologiques, 22 (4), pp. 545–553.

Brunner G.W., 2010. HEC-RAS, River Analysis System, Hydraulic Reference Manual. Davis, CA.: US Army Corps of Engineers, Hydrologic Engineering Center.

Cao Z., Li Y., Yue Z., 2007. Multiple time scales of alluvial rivers carrying suspended sediment and their implications for mathematical modeling. Advances in Water Resources, 30, pp. 715–729.

Cao Z., Hu P., Pender G., 2011. Multiple time scales of fluvial processes with bed load sediment and implications for mathematical modeling. Journal of Hydraulic Engineering, 137 (3), pp. 267–276.

Dysarz T., Wicher-Dysarz J., 2013a. Analysis of flow conditions in the Stare Miasto Reservoir taking into account sediment settling properties. Annual Set the Environmental Protection, 15, pp. 584–606.

Dysarz T., Wicher-Dysarz J., Sojka M., 2013b. Analysis of highway bridge impact on the sediment redistribution along the Stare Miasto reservoir, Poland. Proceedings of 2013 IAHR Congress, Tsinghua University Press, Beijing.

Exner, F.M., 1920. Zur Physik der Dunen. Sitzber. Akad. Wiss Wien, Part IIa, Bd. 129 (in German).

Exner, F.M., 1925. Uber die Wechselwirkung zwischen Wasser und Geschiebe in Flussen. Sitzber. Akad. Wiss Wien, Part IIa, Bd. 134 (in German).

Little C.D., Jonas M., 2010. Sediment Impact Analysis Methods (SIAM): overview of model capabilities, applications, and limitations, 2nd joint Federal Interagency Conference, Las Vegas, NV.

Mooney, D.M., 2006. SIAM, Sediment Impact Analysis Methods, for Evaluating Sedimentation Causes and Effects, Proceedings of the Eighth Federal Interagency Sedimentation Conference, Reno, NV.

Nourani V., 2009. Using artificial neural networks (ANNs) for sediment load forecasting of Talkherood river mounth. Journal of Urban and Environmental Engineering, 3 (1), pp. 1–6.

Papanicolaou A.N., Elhakeem M., Krallis G., Prakash S., Edinger J., 2008. Sediment transport modeling review—current and future developments. Journal of Hydraulic Engineering, 134 (1), pp. 1–14.

Parker G., 2004. 1D Sediment Transport Morphodynamics With Applications To Rivers And Turbidity Currents. e-book published at http://hydrolab.illinois.edu/people/parkerg/.

Rahmanian M.R., Banihashemi M.A., 2011. Sediment distribution pattern in some Iranian dams based on a new empirical reservoir shape function. Lake and Reservoir Management, 27, pp. 245–255.

Salas J.D., Shin H-S., 1999. Uncertainty analysis of reservoir sedimentation. Journal of Hydraulic Engineering, 125 (4), pp. 339–350.

Woliński J., Zgrabczyński J., 2008. The Stare Miasto reservoir in the Powa river: Water management rules, BIPROWODMEL Co., Poznan, (in Polish).

Wu W., 2008. Computational River Dynamics. Taylor & Francis Group, London, UK.

Yang C.T., 1996. Sediment transport theory and practice. McGraw-Hill Series in Water Resources and Environmental Engineering.

Yitian L., Gu R.R., 2003. Modeling Flow and Sediment Transport in a River System Using an Artificial Neural Network. Environmental Management, 31 (1), pp. 122–134.

Reservoir Sedimentation – Schleiss et al. (Eds)
© 2014 Taylor & Francis Group, London, ISBN 978-1-138-02675-9

Influence of geometry shape factor on trapping and flushing efficiencies

S.A. Kantoush
Civil Engineering Program, German University in Cairo, New Cairo City, Cairo, Egypt

A.J. Schleiss
Laboratory of Hydraulic Constructions (LCH), Ecole Polytechnique Fédérale de Lausanne (EPFL), Lausanne, Switzerland

ABSTRACT: The trap efficiency of a shallow reservoir depends on the characteristics of the inflowing sediments and the retention time of the water in the reservoir, which in turn are controlled by the reservoir geometry. With the purpose of controlling the trapped and flushed sediments in shallow reservoirs, the effects of the geometry on sediment deposition and removal were investigated with systematic physical experiments. The geometry shape factor is an important factor to predict the flow and sediment deposition in the reservoir. The evolution of trap efficiency has an increasing or decreasing effect according to the geometry shape factor and flow patterns. The channel formed during flushing attracts the jet and stabilizes the flow structures over the entire reservoir. Empirical formulas to describe the relationship between the geometry shape factor and sediment trap efficiency as well as flushing efficiency were developed.

1 INTRODUCTION

1.1 *Problematic of sedimentation in shallow reservoirs*

Suspended sediment deposition is a complex phenomenon in deep and shallow reservoirs. The sediment deposition in shallow reservoirs of run-of-river power plants reduces the storage capacity and generates a risk of blockage of intake structures as well as sediment entrainment in hydropower schemes. The planning and design of shallow reservoir require the accurate prediction of sediment trapping and release efficiencies. Reservoir sedimentation is the principal cause preventing sustainable use of storage reservoirs (Annandale, 2013). Many flows in nature can be considered as shallow as for examples flows in wide rivers, lakes, bays and coastal regions. For such flows, the horizontal dimensions are much larger than the vertical depth and turbulence is of a special nature, as was discussed already by Yuce and Chen (2003). Reservoir sedimentation has been methodically studied since 1930s (Eakin, 1939), but dam engineering has focused on structural issues, giving relatively little attention to the problem of sediment accumulation (De Cesare and Lafitte, 2007). The problem confronting the designer is to estimate the rate of deposition and the period of time before the sediment will interfere with the useful functioning of a reservoir. Several concepts of reservoir life may be defined as its useful, economic, useable, design and full life as adapted from (Murthy, 1977), (Sloff, 1991). The rapid reservoir sedimentation not only decreases the storage capacity, but also increases the probability of flood inundation in the upstream reaches due to heightening of the bed elevations at the upstream end of the reservoir and the confluences of the tributaries (Liu et al., 2004). In order to remove and reduce reservoir sedimentation, many approaches such as flushing, sluicing, dredging and water and soil conservation are developed (ICOLD, 1989). Among these approaches, flushing is considered the only economic approach to swiftly restore the storage capacity of

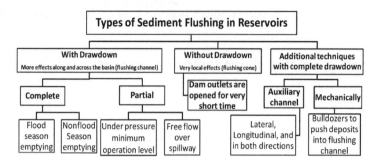

Figure 1. Classification of sediment flushing techniques.

the reservoir with severe deposition. Basically, there are two types of flushing operations with, and without drawdown, and optional techniques can be used with the complete drawn flushing as shown in Figure 1.

1.2 *Reservoir trap efficiency*

Brune's curve is used by dam engineers to predict the trap efficiency which is the proportion of deposited to flowing sediments into a reservoir (Brune, 1953). This curve provides an estimate of the relative amounts of sediment that will be retained by reservoirs of various sizes. The size of a reservoir is quantified by using the same measure we defined before. Recall that we indicated that a good way of determining the relative size of a reservoir is to divide its storage volume by the mean annual river flow (Annandale, 2013). Churchill (1948) based his empirical relationship on the concept of sediment releasing, whereas Brune (1953) used the concept of sediment trapping which has come into more common use. Several approaches have been undertaken to quantify sediment trap efficiency. Churchill (1948) presented a curve relating the trap efficiency to the ratio between the water retention time and mean velocity in the reservoir. As a result of the complexity of the phenomenon involved in sediment deposition in lakes and reservoirs, focused research efforts on numerical and laboratory modeling also have been published. The Trap Efficiency (*TE*) of reservoirs depends on several parameters (an overview of the processes taking place in a reservoir is given by (Heinemann, 1984). Since *TE* is dependent on the amount of sediment, parameters controlling the sedimentation process are shown in Figure 2.

Therefore, the particle-size distribution of the incoming sediment controls TE in relation to retention time. Coarser material will have a higher settling velocity, and less time is required for its deposition. Very fine material, on the other hand, will need long retention times to deposit. The retention time of a reservoir is related to: 1. The characteristics of the inflow hydrograph and; 2. The geometry of the reservoir, including storage capacity, shape and outlet typology. The reservoir geometry can also govern the retention time.

1.3 *Empirical and theoretical models for predicting TE*

Simple models relating TE to a single reservoir parameter are, on the other hand, easy to implement but are far less accurate. One has to distinguish between the TE of a reservoir on a mid to long-term basis and its TE for one single event. Heinemann (1984) gave an overview of the many empirical models that could be used for predicting TE. An overview of the theoretically based TE models is provided by Haan et al. (1994). Verstraeten and Poesen (2000) provided an overview of the different methods available to estimate the trap efficiency of reservoirs and ponds. As already mentioned the empirical models predict trap efficiency, mostly of normally ponded large reservoirs using data on a mid to long-term basis. These models relate trap efficiency to a capacity/catchment ratio, a capacity/annual inflow ratio or a sedimentation index. Today, these empirical models are the most widely used models to predict trap efficiency, even for reservoirs or ponds that have totally different characteristics from the

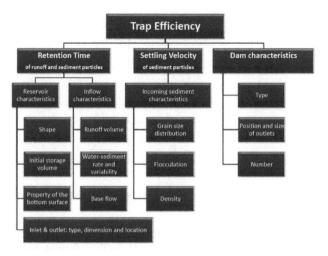

Figure 2.   Factors influencing the trap efficiency of reservoirs.

reservoirs used in these models. For small ponds, these models seem to be less appropriate. Furthermore, they also cannot be used for predicting trap efficiency for a single event.

At present only limited research has been done on establishing mid-term trap efficiency models based on theoretical principles. This is probably the most important gap in trap efficiency research that the present study is presenting. The study aims at investigating the evolution of the sediment trap efficiency based on ten laboratory test with various reservoir geometries, and proposes measures to control and predict the channel characteristics. Therefore this study is focusing mainly on influence of the reservoir geometrical parameters as Aspect Ratio AR, Expansion Ratio ER, Expansion area ratio σ, Expansion density ratio and geometry shape factor SK, on trap and flushing efficiencies. Finally, several empirical formulas that describe the relationship between the reservoir geometry and sediment trap efficiency as well as flushing efficiency for two modes of flushing were developed. An empirical formula for drawdown flushing describes the function between the geometry shape factor and flushing efficiency is presented.

2   PHYSICAL EXPERIMENTS

2.1   *Experimental facility*

The experimental tests have been conducted in a rectangular shallow basin with inner maximum dimensions of 6.0 m in length and 4 m in width, as sketched in Figure 3(a). The inlet and outlet rectangular channels are both 0.25 m wide and 1.0 m long. The bottom of the basin is flat and consists in hydraulically smooth PVC plates. The walls, also in PVC, can be moved to modify the geometry of the basin. Adjacent to the reservoir, a mixing tank is used to prepare the water-sediments mixture. The water-sediments mixture is supplied by gravity into the water-filled rectangular basin. Along the basin side walls, a 4.0 m long, movable frame is mounted to carry the measuring instruments. The sediments were added to the mixing tank during the tests. To model suspended sediment currents in the laboratory model, walnut crushed shells with a median grain size $d_{50} = 50$ μm, density 1500 kg/m³ was used in all experiments. These are non-cohesive, light weight and homogeneous grain material. The bed level evolution was measured with a Miniature echo sounder (UWS). The sounder was mounted on a movable frame which allowing to scan the whole basin area. The sediment concentrations of the suspensions material using the crushed walnut shells were measured. The hydraulic and sediment conditions were chosen to fulfill the sediment transport requirements. Furthermore, for all tests, Froude number ($0.05 \le Fr \le 0.43$) was small enough and Reynolds

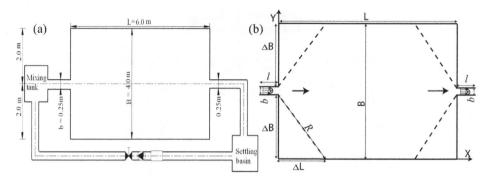

Figure 3.　(a) Plan view of the laboratory setup; (b) Geometrical parameters of the test configurations.

Table 1.　Configurations of different test series and their geometrical characteristics: L and B are length and width, A the total surface area of the basin, ER and AR are the expansion and aspect ratios, P is the wetted perimeter of the length of the side walls, and SK is the shape factor $SK = (P/\sqrt{A})*AR*D_{exp}$.

Test	B [m]	L [m]	P [m]	A [m²]	$AR = L/B$ [-]	$ER = B/b$ [-]	$SK = (P/\sqrt{A})*AR*D_{exp}$ [-]	Form
T1, T2, T3, T4	4.0	6.0	19.5	16	1.5	16	5.97	
T7	3.0	6.0	17.5	12	2.0	12	8.25	
T8	2.0	6.0	15.5	8	3.0	8	13.42	
T9	1.0	6.0	13.5	4	6.0	4	33.07	
T11	4.0	5.0	17.5	16	1.25	16	4.89	
T12	4.0	4.0	15.5	16	1.0	16	3.88	
T13	4.0	3.0	13.5	16	0.75	16	2.92	
T14	4	6	14.1	8	1.5	8	11.2	
T15	4	6		16	1.5	16	11.08	
T16	4	6		12	1.5	25	4.65	

number ($14000 \le Re \le 28000$) high enough to ensure subcritical, fully developed turbulent flow conditions.

2.2　Test configurations

Ten axi-symmetric basins with different forms were tested to study the geometry shape effect on the flow and deposition pattern (see Table 1). In order to gain insight into the physical process behind the sedimentation of shallow reservoirs governed by suspended sediment; a reference basin geometry with width of B = 4.0 m and length of L = 6.0 m was used. The reference

132

geometry was used for the first six tests, from Test 1 (T1) to Test 6 (T6), to examine different test procedures and find the optimal one to continue with future test configurations. As a reference case, the rectangular basin geometry was analyzed in detail (Kantoush, 2008, Kantoush and Schleiss, 2010). To investigate the effect of the basin width effect on the flow and sedimentation processes in the reservoir the experiments focused on the width achieved in rectangular reservoir 6.0 m long and 3.0, 2.0, 1.0, 0.5 m wide (from T7 to T10), respectively. With a second set of tests the effect of the basin length experimental tests have been conducted in a rectangular shallow basin 4.0 m wide and 5.0, 4.0, and 3.0 m long (from T11 to T13), respectively. Finally geometries with three expansion angles were tested (from T14 to T16). In the present paper the results of flushing for experiments T1, T8, T14, and T16 are presented hereafter.

2.3 Geometrical parameters

The geometrical parameters are defined in Figure 3(b) and all tests are summarized at Table 1. In order to represent all geometrical characteristics parameters with flow and deposition results, a geometry shape factor SK was developed. In the present study several reservoir geometries with different shapes have been conducted. Thus, there is a need for a dimensionless coefficient representative of different geometry shapes which can be correlated with flushing efficiency. The following definitions are used (see Figure 3(b) & Table 1):

- Length and the width of the upstream and downstream channels which remained constant for all configurations: $l = 1.0\ m$, $b = 0.25\ m$ and $l = 4\ b$
- Length and the width of the basin: L and B;
- Depth of lateral expansion ΔB;
- Distance from the edge of channel to the edge of the basin R;
- Total surface area of the basin A;
- Lateral expansion ratio: $ER = B/b$;
- Aspect ratio as $AR = L/B$;
- Jet expansion density can be defined as $D_{exp} = R/\Delta B$;
- Geometry shape factor can be defined as $SK = (P/\sqrt{A})*AR*D_{exp}$.

2.4 Test procedure

After filling the basin and having reached a stable state with the clear water. First LSPIV recording (Large Scale Particle Image Velocimetry) has been performed during 3 minutes. Then a second phase, the water-sediment mixture was drained by gravity into the water-filled rectangular basin. The flow circulation pattern with suspended sediment inflow was examined every 30 minutes using LSPIV over 90 minute's period. The flap gate was then closed to permit for the suspended sediment to deposit and then start bed level profile measurements by using UWS. Every 1.5 hrs, the bed morphology was measured at different cross sections. After each time step the pump was interrupted to allow bed morphology recording. The final bed morphology was used as the initial topography for two modes of flushing (free flow and drawdown flushing). Clear water without sediment was introduced into the basin to investigate the effect of free flow and drawdown flushing. Normal water depth of h = 0.20 m was used without lowering of reservoir during free flow flushing. With lowering the water depth in the reservoir to half of the normal water depth (h = 0.10 m), the drawdown flushing was conducted. Each mode of flushing lasted for two days with flow field and final bed morphology measurements.

3 RESULTS AND DISCUSSIONS

3.1 Influence of Aspect Ratio of reservoir (AR) on Trap Efficiency (TE)

Various geometrical configurations with different aspect ratio $AR = L/B$, where L and B is length and width of the reservoir, respectively, hydraulic and sediment conditions have been analyzed. By knowing the actual deposited sediment (V_{dep}) and the sediment volume flowing into a reservoir (V_{in}), it is possible to calculated the V_{dep}/V_{in} ratio and determine the percentage of sediment flowing into a reservoir that will be trapped *TE*. Figure 4(a) shows the trap efficiency *TE* as a function of the reservoir aspect ratio, AR, at six measurement periods (from t_1 to t_6). The trap efficiency is ranging from 98% at t_6 of 1080 minutes ($AR = 1.5$) to 38% at t_1 of 90 minutes ($AR = 0.75$). The TE has a rising tendency while *AR* increasing until reaches a critical *AR* value. Then *TE* starts to decrease for a higher *AR*. The *TE* curves at $t_2 = 180$ min and $t_3 = 270$ min have approximately the same trend as for t_1. That means the reservoirs did not reach to equilibrium state and flow patterns were changing during these three periods. The reservoir reached to a quasi-equilibrium state during the longest test duration of $t_6 = 1080$ min, since the observed *TE* reached to 100%. Equilibrium is associated with vanishing the cumulative net of sediment concentration but this does not imply that the instantaneous sediment flux vanishes. It can be concluded that *TE* increases with increasing reservoir aspect ratio until it reached the highest *TE* and then it decreases with increasing aspect ratio as shown in Figure 4. Lesser amounts of sediment may be retained by reduced aspect ratio of reservoir.

3.2 *Influence of Expansion Ratio (ER) of reservoir on Trap Efficiency (TE)*

The influence of the Expansion Ratio (ER) on the trap efficiency TE is illustrated in Figure 4(b). It can be seen that the increase of the efficiency between ER of 8 and 12 is the same order of magnitude as the increase between 16 and 26 (about 40%–45%). Therefore, the changes from an asymmetric flow pattern to a symmetric flow pattern for ER of 16 are responsible for a break in the efficiency curve of the reservoir. The evolution of the trap efficiency is compared for six different runs at 90, 180, 270, 450, 540, and 1080 minutes. Several data points are located at ER = 16 which indicated that expansion ratio is not representative for geometries with a fixed width and variable length. Trap efficiency increases with increasing ER till it reaches ER = 12 where TE is almost 100%. The minimum TE was obtained for a

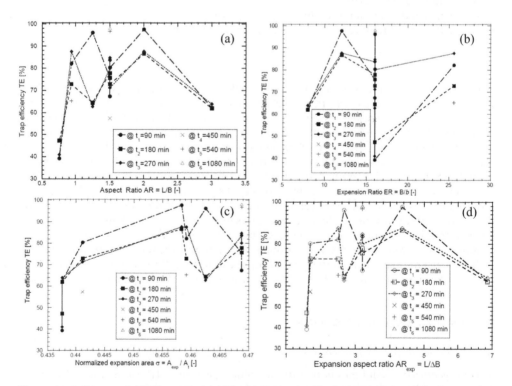

Figure 4. Influence of (a) Aspect Ratio (*AR*), (b) Expansion Ratio (*ER*) (c) normalized expansion ratio, (d) Expansion Aspect Ratio AR$_{exp}$, on trap efficiency of reservoir $TE = V_{dep}/V_{in}$.

basin with ER = 16. For higher ER, the trap efficiency decreases again. There is no significant changes of TE evolution for narrow reservoirs with low of ER = 8.

3.3 Influence of normalized expansion area $\sigma = A_{exp}/A_t$ of reservoir on Trap Efficiency (TE)

The effect of expansion area ratio is defined as ratio of expansion to total surface areas. Figure 4(c) shows the influence of σ on the TE. In beginning after first period $t_1 = 90$ min, the trap efficiency increases with rising expansion area ratio σ until a maximum TE value is reached. Then, it decreases again for higher expansion area ratio. Almost the same trend was found at t_2 and t_3.

3.4 Influence of normalized expansion aspect ratio $AR_{exp} = L/\Delta B$ of reservoir on TE

The evolution of trap efficiency as a function of expansion aspect ratio $AR_{exp} = L/\Delta B$ was investigated as shown in Figure 4(d). Trap efficiency is rising for higher expansion aspect ratio, which can be defined as the ratio of expansion length to the expansion width. It reaches a maximum TE at $AR_{exp} = 3.2$ before it decreases again with for higher AR_{exp}. The trap efficiency after $t_1 = 90$ min reached to the peak; afterwards it declined by almost 15%. Finally TE increased again to asymptotically reaching 100% before decreasing again with further increase of AR_{exp}.

3.5 Influence of geometry shape factor SK on trap efficiency TE

There are several non-dimensional geometrical parameters as (AR, ER, σ, and A_{exp}) that have no clear influence on trap efficiency by considering each parameter separately. Therefore, a set of several combinations of these parameters were used and analyzed versus trap efficiency. It was found that geometry shape factor SK affects the trap efficiency. The geometry shape factor SK is defined as the $SK = (P/\sqrt{A}) * AR * D_{exp}$. The evolution of trap efficiency and relationship with geometry shape factor was depicted in Figure 5(a). It is clearly visible in that TE decreases with increasing geometry shape factor (SK). The evolution of trap efficiency is increased with time and it reached quasi equilibrium during the last period of last run. An empirical relationship between trap efficiency TE in percentage and geometry shape factor SK was developed from all experiments in Eq. (1) with application range of $2.92 < SK < 13.42$.

$$TE = V_{dep}/V_{in} = 95 - 360 \ (SK)^{-2} - 12 \ (SK/10)^2 \tag{1}$$

It seems that smaller geometry Shape factor SK trapped less sediments and the evolution of trap efficiency can be approximated by a fitting decreasing curve as shown in Figure 5(a) for SK = 2.92. It can be conclude that the distance between inlet and outlet of the reservoir has a great influence of trap efficiency. By increasing SK to 3.41 trap efficiency increased by almost 35% and the evolution curve with decreasing tendency at the end of the experiment. With further increasing for SK still trap efficiency increases as shown in Figure 5(a). It is clearly visible in Figure 5(a) that the evolution tendency of TE decreases again from SK > 10, as the flow pattern was straight from inlet to outlet with no or one circulation cells inside the reservoirs. Therefore, the minimum deposited volume was obtained for higher geometry shape factor SK > 10. It can be concluded that flow pattern with no/odd number of cells are preferable to reduce depositions.

3.6 Prediction of the drawdown flushing efficiency

Efficiency of flushing of suspended sediment through the reservoir is important to determine the feasibility of flushing operations according to the designed reservoir. The measured data for each run were recorded after one time of a flushing with clear water was performed during two days. With the total cumulative deposited sediment at the end of each experiment and volume of flushed sediments during this procedure, flushing efficiency, FE, is defined as:

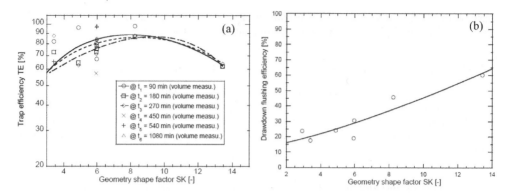

Figure 5. Influence of geometry shape factor SK on (a) trap efficiency; (b) flushing efficiency.

$$FE = V_{flushed}/V_{df} \qquad (2)$$

where $V_{flushed}$ is the volume of flushed sediment with clear water after two days, V_{df} is the total cumulative deposited volume after a specific period. Figure 5(b) shows the influence of the geometry shape factor SK on drawdown flushing efficiency FE. Flushing efficiency FE is an index to describe the effectiveness of hydraulic flushing. With lowering the water depth in the reservoir to half of the normal water depth, the efficient drawdown flushing as a function of geometry shape factor were plotted in Figure 5(b). It was observed that the drawdown flushing efficiency increases with higher geometry shape factor. The drawdown flushing was higher compared to the free flow flushing. The minimum flushing efficiency was 20% for the smallest geometry shape factor. Additionally, the maximum drawdown flushing efficiency reached almost 65% for the highest geometry shape factor as shown in Figure 5(b). It was found qualitatively that almost fifty percent of the total volume removed sediments were flushed out in the one fourth of the flushing duration. Flushing efficiency was correlated with the geometry shape factor the empirical relationship formula in Eq. 3. The application range of Eq. 3 is 2.92 < SK < 13.42 and the coefficient of determination $R^2 = 0.92$.

$$FE = 103 + 12.4 \cdot (SK/10)^{-2} - 65.75 \cdot (SK/10)^{-1} \qquad (3)$$

where FE is flushing efficiency ($FE = V_{flushed}/V_{df}$) and SK is the geometry shape factor ($SK = (P/\sqrt{A})*AR*D_{exp}$).

4 CONCLUSION

The trap efficiency reduces with time for the lozenge form (T14, SK = 11.2), but TE is high in the beginning. With the lozenge form the jet could expanded over all the basin geometry. According to the proposed empirical formula, trap efficiency can be estimated. The length of the reservoir plays a critical role in determining the jet flow type and the associated pattern of the bed deposition. The use of an elongated basin increases the retention of sediments as it is well known for sand trap basins. The maximum Aspect Ratio (Length to width ratio) of reservoir should be 1.5. For the experiments with drawdown flushing it is important to know the channel width and length in order to estimate the gain of the reservoir capacity. Due to the sensitivity of the flow pattern on the boundary conditions, initial conditions and the geometry (and changes in time), it is difficult to predicate the exact location of the flushing channel. Drawdown flushing efficiency becomes better for increasing geometry shape factor SK as given by the empirical relationship in Eq. 3 within the application range of is 2.92 < SK < 13.42.

REFERENCES

Annandle, G, 2013. Quenching the Thirst: Sustainable Water Supply and Climate Change. ISBN: 1480265152, *Library of Congress Control Number: 2012921241, CreatSpace Independent Publishing Platform*, North Charleston, SC.

Brune, G.M. 1953. Trap efficiency of reservoirs. *Transactions of Americ. Geophys. Union*, 34:407–418.

Churchill, M.A., 1948. Discussion of analysis and use of reservoir sedimentation data.

De Cesare, G. and Lafitte, R, 2007. Outline of the historical development regarding reservoir sedimentation. 32nd *Congress of IAHR, Harmonizing the Demands of Art and Nature in Hydraulics*, Venice.

Eakin, H.M., 1939. Instructions for reservoir sedimentation surveys, in silting of reservoirs. *Technical report, U.S. Department of Agriculture*, Technical Bulletin.

Haan, C.T., Barfield, B.J., and Hayes, J.C., 1994. Design hydrology and sedimentology for small catchments. *San Diego*, Academic Press.

Heinemann, H.G., 1984. Reservoir trap efficiency. Erosion and sediment yield: some methods of measurement and modelling, *Norwich: GeoBooks*, Norwich.

ICOLD. 1989. Sedimentation control of reservoirs. Bulletin of International Committee of Large Dams.

Kantoush, S.A. and Schleiss, A.J., 2009. Channel formation in large shallow reservoirs with different geometries during flushing. *Journal of Environmental Technology*, Volume 30 Issue 8, 855–863.

Kantoush, S.A., 2008. Experimental study on the influence of the geometry of shallow reservoirs on flow patterns and sedimentation by suspended sediments, EPFL Thesis No. 4048 and Communication No. 37 of Laboratory of Hydraulic Constructions (LCH), EPFL, ISSN 1661–1179.

Murthy, B.N.,1977. Life of reservoir. Central Board of Irrigation and Power.

Sloff, C.J., 1991. Reservoir sedimentation: a literature survey. Technical report, *Communications on Hydraulic and Geotechnical Engineering Faculty of Civil Engineering*, Delft University of Technology.

Yuce, M.I. and Chen, D, 2003. An experimental investigation of pollutant mixing and trapping in shallow costal re-circulating flows. *In Proc. the Int. Symp. on shallow flows*, Part I:165–172, Delft, The Netherlands.

Flushing of coarse and graded sediments—a case study using reduced scale model

J.R.M. Almeida, J.J. Ota, F.R. Terabe & I.I. Muller
Lactec Cehpar—Instituto de Pesquisa para o Desenvolvimento, Curitiba, Paraná, Brazil

ABSTRACT: The aim of this paper is to present an experience of coarse sediment draw-down flushing tests carried out in a reduced scale model for Palomino Hydroelectric Project (Dominican Republic). It characterizes the volume of reservoir recovered over time after the opening of the bottom outlet gate. Tests were performed with a movable bed composed by non-cohesive graded material. Its gradation was controlled according to the Froude similar-ity in term of the shear stresses that cause the entrainment according to Shields criterion. The purged volume was measured by the sediment concentration of released discharge and by the topographic measurement in the reservoir and both results achieved similar values. The model tests proved to be highly appropriate for showing the region where the volume recovery is more intense, i.e. whether flushing is really useful for the reservoir.

1 INTRODUCTION

Problems of sediment accumulation in reservoirs built on natural beds have been recurrent in engineering. For this reason structures whose purpose is to enable the transfer of accu-mulated sediments to the region downstream from the developments are constantly foreseen in the projects. This operation is called flushing of sediments, and it has not always achieved its purpose. Considering the specific type of sediment and the importance of recovering reservoir useful storage volume, it became necessary to perform a detailed study of flushing for the Palomino Hydroelectric Plant reservoir. It is known that considerable drawdown of the water level must be performed, opening the bottom outlet gate to create a strong energy gradient sufficient to entrain the reservoir sediment.

Several factors determine the efficiency of these structures. Outstanding among them are topography and geomorphology of the regions, the hydraulic characteristics of the river, the characteristics of the alluvial material to be purged and, obviously, the discharge structure and how it is operated. As seen in Chella (2002), LACTEC/CEHPAR (Brazil) has devel-oped tests with movable bed in reservoirs and considers that it is very useful to simulate these operations in reduced scale models. In these models it is possible to evaluate the efficiency of the structures, test several alternatives and also optimize the operation of the sediment discharge structures, enabling more rational operation and increasing the useful life of the reservoir.

2 CHARACTERISTICS OF THE PALOMINO HYDROELECTRIC PROJECT

The Palomino Hydroelectric Project aims to generate energy by diverting part of the flow of the Yaque del Sur and Blanco rivers to the Rio del Medio river. The installed power of the plant is 80.0 MW, generated by Francis type turbines associated with a gross head of approxi-mately 335.0 m. The reservoir level of water for generation ranges from El. 790.0 m (elevation of the spillway crest without gates) to El. 784.0 m. The water intake is located close to the left bank of the dam, at El. 773.5 m. Plant operation is foreseen only for the period with the highest energy demand in the day, with a water consumption higher than the inflow to the reservoir. Therefore, in order to allow generation, it is necessary to accumulate water during the time of day when no generation occurs. For plant operation, it is necessary to maintain a volume that can store this water, and it is not desirable to have the presence of sediment that will diminish the live storage space of the reservoir.

At the water intake a bottom outlet was planned to enable the transfer of the solid material deposited in the reservoir towards the area downstream of the dam. The bottom outlet is controlled by a tainter gate, 6.0 m wide by 6.5 m high. It must be able to reduce the water level in the reservoir and carry out flushing. When the water level in the reservoir is at its highest (El. 790.0 m) the bottom outlet can release 680 m³/s with completely open gate.

3 REDUCED SCALE MODEL STUDIES

The tests to evaluate bottom outlet efficiency in removing accumulated sediments were performed in the reduced scale model with a geometric scale of 1:70 and operated according to the Froude similarity criterion.

3.1 *Reproduction of alluvial material*

Many studies of sediment transport in reduced scale models are performed with special solid material with a greater diameter and smaller density, as cited in Novak and Cabelka (1981). The CEHPAR has performed sedimentation and entrainment studies in reduced scale models reproducing the alluvial material using treated imbuia wood sawdust (Chella, 2002), with a density of 1150 kg/m³ to simulate sand particles (density 2650 kg/m³ with a fine diameter and homogeneous). However, the alluvial material in the area of this enterprise, indicated in Figure 1, is well graded and has significant percentage of large grains (up to 0.8 m in diameter). For cases like this, it is not appropriate to use imbuia sawdust. Due to its uniformity, it is impossible to reproduce in the model the occurrence of segregation of fine and coarse fractions of the material, a process known as armoring.

Therefore it was decided to use a non-cohesive granular material, composed by sand and gravel, with a grain size distribution selected appropriately. The grain size of the material used was determined in such a way as to allow similarity in the critical shear stresses that provoke its entrainment. Thus, the material was specific so that each grain size range would be entrained when the grains that composed it were submitted, in the model, to shear stresses similar to those that would be observed in the prototype if these particles were moved.

In order to determine this similarity, the simple application of the Froude criterion was not adopted. This would reduce the grain size curve of the prototype geometrically on the scale of 1:70. The methodology used was as follows: with the diameter of each range of the material, in the prototype, sedimentation velocity ω and critical shear velocity v^*_{cr} were determined.

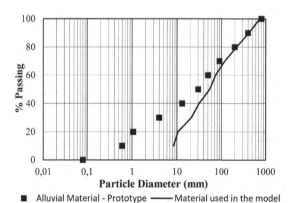

Figure 1.   Grain size analysis to determine the material to be used in the tests.

The scale of velocities (Froude similarity) was applied to these values to obtain the velocities ω and v^*_{cr} in the model. Based on these values, the diameters of the particles in the model were obtained, which would be similar from the point of view of entrainment capacity. The settling velocity (ω) was determined using the classical figure that presents settling velocity according to the diameter and the shape factor at different temperatures—U.S. Inter-Agency Committee on Water Resources, Subcommittee on Sedimentation (1957). A shape factor equal to 0.7 was used, and a temperature of 20° C. For the critical shear velocity (v^*_{cr}) the well known Shields criterion was used. The grain size resulting from the material used in the model is represented in Figure 1. To make the comparison easier, all grain size curves are in prototype magnitudes. It is observed that the results to adopt similarity of settling velocity and shear velocity were practically the same, slightly larger for the smaller diameter range compared to simple geometric transformation of the material.

This criterion used to determine the grain size of the material is certainly more effective than the simple geometrical transformation, since it reproduces the physical mechanisms that rule sediment transport in canals most correctly. It enabled utilizing a coarser material that is less influenced by the reduced scale, since it is not easily transported by suspension in water—a less important process for flushing material from the Yaque del Sur river which contains coarse material subjected to entrainment movements.

3.2 Tests performed

In order to evaluate the efficiency of the bottom outlet to remove sediment accumulated in the reservoir, 6 tests were carried out, with a duration equivalent to 12 hours and 30 minutes, simulating the reservoir purging operation for two different initial sedimentation levels: 784.0 m and 790.0 m (total silting up until the spillway crest) and three inflows, as indicated in Table 1. The purging operations were performed with complete opening of the bottom outlet tainter gate.

In these tests outflowing water was collected from the bottom outlet every 15 minutes to determine the concentration of solid material present in the jet (C_s). With this information it was possible to obtain the volume of reservoir recovered over time, by Equations 1 and 2, and thus evaluate the efficiency of the purging operation. Topo bathymetric surveys of the mobile bed upstream from the dam were also performed after 4 hours of test duration and at the end of test in cross sections with a 40 m spacing.

$$Q_S = C_S \cdot Q \tag{1}$$

$$V_{rec} = \frac{\rho_s}{\rho} \cdot \int Q_S \cdot dt \tag{2}$$

where: Q_s = solid discharge of sediments; C_s = volume concentration of sediments in the jet discharged from the bottom outlet; Q = outflow from the bottom outlet adopted as being equal to the inflow into the reservoir; V_{rec} = reservoir volume recovered by the purging operation; ρ_s = density of solids; and ρ = density of water.

Table 1. Initial conditions of the tests performed.

Nr. of test	Total inflow (m³/s)	Level of sediments in the reservoir (m)
1	50.0	784.0
2	75.0	784.0
3	150.0	784.0
4	50.0	790.0
5	75.0	790.0
6	150.0	790.0

3.3 Results obtained

The material was entrained through canals that formed in the mobile material bed initially imposed in the reservoir. It was observed that these canals generally had trapezoidal sections, and over time the geometry varied greatly, which is evident that the sediment transport process is not continuous. However, common to all the tests was that at some moment armoring occurred in the more distant regions of the dam, as indicated in Figure 2, which limited the recovery of the reservoir volume due to the diminished entrainment of sediment in these regions.

According to White (2001), the embankment slopes in the trapezoidal section (i.e. of the banks) resulting from the purging operation are steeper, and may even be almost vertical, the more consolidated are the deposits in the reservoir. In reduced scale model tests, the consolidation of the sediment deposit cannot be reproduced. However, it was observed that, in certain cases, the slopes of the canal embankments were practically vertical due to the cohesive force among the wet sand particles, which in a way makes the model results slightly conservative. But this process provokes the collapse of the bank that is in a non regular form, thus, the sediment concentration may suddenly increase after a bank slumping.

Figure 3 presents the results obtained in the readings of the concentration under the two extreme conditions tested: Test 1—smaller discharge and less sedimentation. Test 6—greater discharge and greater sedimentation. Since the sediment transport process is not continuous and permanent, there is a great variation of the amount of solid material concentration present in the jet discharged from the bottom outlet throughout the test. In order to improve

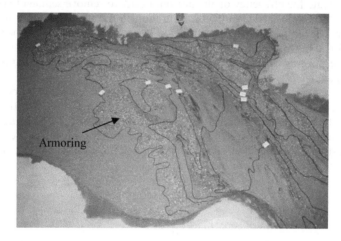

Figure 2.   Armoring in the regions that are further away from the dam—Test 6.

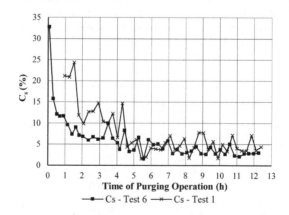

Figure 3.   Concentration of solid material in the bottom outlet jet—Tests 1 and 6.

understanding of the evolution of purging operation efficiency, based on the derivation of a polynomial equation that reproduces the volume recovery of the reservoir, a curve of the concentration tendency of solid material present in the jet discharged from the bottom outlet was adjusted. Figures 4 and 5 show, for instance, the adjustment performed for Test 6.

Figure 6 shows the concentrations of solid material in the bottom outlet jet for all tests. This allows evaluating the efficiency of the process without taking into account the

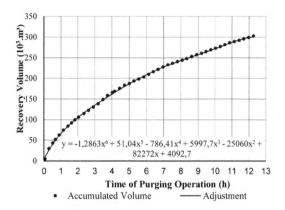

Figure 4. Test 6—Adjustment for the recovered reservoir volume.

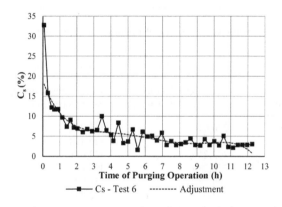

Figure 5. Test 6—Adjustment for the evolution of the concentration of solid material in the bottom outlet jet.

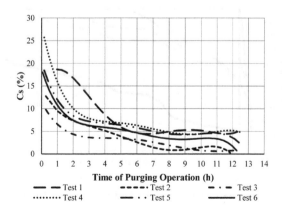

Figure 6. Concentration of solid material present in the bottom outlet outflow jet.

instantaneous values of C_s which are highly influenced by the changes in the conditions of material entrainments and collapse of the banks. It is observed that, in all situations, the value of C_s tends to stabilize at around 5% after 5 hours of purging operations. Figure 7 shows the reservoir volume recovery curves throughout the tests.

The evaluation of the reservoir volume recovered based on the information about concentrations of solid material in the bottom outlet jet is adequate since, as indicated in Table 2, the comparison of the final volume calculation was similar to that obtained from the topo bathymetric surveys.

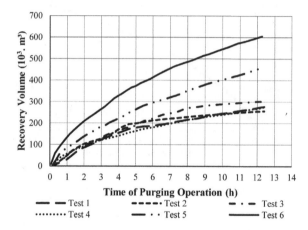

Figure 7. Relation of reservoir recovery x time of operation and purging.

Table 2. Comparison of the reservoir volume recovered.

Nr. of test	Volume based on the data of C_s (m³/s)	Volume based on topo bathymetry (m³/s)
1	276,646	285,964
2	256,827	320,488
3	302,240	372,611
4	278,280	398,719
5	460,532	531,905
6	604,677	732,490

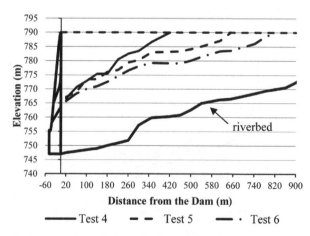

Figure 8. Longitudinal profile of canals formed—Tests 4, 5 and 6.

White (2001) suggests, for the prior evaluation of the reservoir volume recovered by the purging operation, that the canal geometry be simplified by a trapezoidal cross section and constant slope, similar to the natural slope of the river. Figure 8 shows the profiles of the canals formed by purging operations for tests on initial sedimentation level at El. 790.0 m. The mean slopes observed were 5.7% for Test 4, 3.1% for Test 5 and 2.6% for Test 6. Figure 9 shows the final configuration of the sediment deposit after the end of Test 6. It is noted that for the case studied, the simplification was confirmed only for cases of purging operations with the 75.0 and 150.0 m³/s flows. It is interesting to compare these slopes with Equation 3, obtained by Pinto (1977) for cofferdam (diversion sill) built in flowing water by normal dump river closure. This equation assumes Shields critical condition for relatively large sediments and Manning equation with Strickler coefficient for a flume in equilibrium with uniform flow with specific discharge q.

$$i = 0,245 \cdot \left(\frac{\rho_s - \rho}{\rho} \right)^{10/7} \cdot \left(\frac{D^{9/7}}{q^{6/7}} \right) \tag{3}$$

where: ρ_s = specific density of solid grains (equal to 2650 kg/m³ for the tested material); ρ = specific density of water (equal to 1000 kg/m³); D = representative diameter of the material used in the model (having adopted D_{75}); q = specific flow of model over mobile bed.

As has already been discussed, the geometry of the canals is very variable during the test. For this reason it is difficult to know precisely the value of the specific flow. For this study, based on observations made during the tests, a mean width was adopted equal to approximately 23.0 m. It should be emphasized that the discharge used for this calculation is equal half the test discharge, since the reservoir is formed in two rivers, and the condition tested is distribution of 50% in each of them.

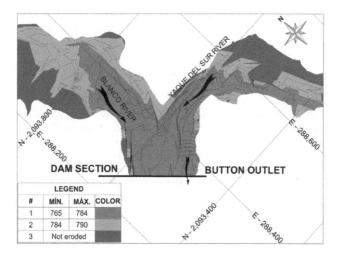

Figure 9.   Configuration of sediment deposit at the end of Test 6.

Table 3.   Theoretical (Eq. 3) and experimental slopes for tests 4, 5 and 6.

Nr. of test	Theoretical slope (%)	Experimental slope (%)
4	5.5	5.7
5	3.1	3.1
6	2.2	2.6

The diameter used for calculations was D_{75}. This choice is justified because the coarse particles are more important in forming the armor that protects the bed, since the fines are easily removed.

Table 3 summarizes the calculations (Equation 3) made for the previously described tests. It also highlights the deviations found among the experimental and theoretical slopes.

4 CONCLUSION

Traditionally, sedimentation and entrainment tests in reduced scale models are carried out reproducing the alluvial material with special materials, such as sawdust, bakelite, coal, etc. For the Palomino Project study, however, a non-cohesive granular material was used, with a grain size distribution selected so that there would be similar critical shear stresses that provoke its entrainment. The material selected proved adequate, since it was possible to reproduce the physical mechanisms, including the onset of armoring, which rule the transport of coarser sediments, such as those that exist in the region of the project.

Similarly, the volume transported in movable bed tests on reduced scale models is usually defined only by topo bathymetric surveys in well defined step of the test. However, the studies performed sought to characterize the sediment purging operation over time by measuring the sediment concentration of the flow from the bottom outlet. This methodology proved adequate, since the volumes obtained from the information about solid material concentration in the jet of the bottom outlet were similar to those obtained from the topo bathymetric surveys.

The tests adopted a few simplifications such as constant flow, homogeneous sedimentation material throughout the reservoir and initial sedimentation defined by a plane on a specific level. But certainly, the physical modeling is able to supply important information taking into account the different particularities of the reservoir configuration, the layout of the structures and the bottom outlet. The identification of the reservoir region where volume was recovered was very useful to confirm that the flushing process enables the recovery of a large part of the useful storage volume of the reservoir.

REFERENCES

Chella, M.R. 2002. *Physical Simulation of Sediment Transport and Sedimentation in Reservoirs—a Case Study for Melissa Hydroelectric Development. Master Thesis.* Curitiba: Parana Federal University. (in Portuguese).

Pinto, N.L.S. 1977. *Contribution to the Study of Rockfill Dams Constructed in Flowing Water.* Curitiba: Parana Federal University (in Portuguese).

Straub, L.G. 1963. *Caroni River Hydroelectric Development. Report on Gury Project River Diversion Scheme.* St. Anthony Falls Hydraulic Laboratory.

White, R. 2001. *Evacuation of Sediments from Reservoirs.* Bristol, UK. HR Wallingford: Ed. Thomas Telford Publishing.

Reservoir Sedimentation – Schleiss et al. (Eds)
© *2014 Taylor & Francis Group, London, ISBN 978-1-138-02675-9*

1D modelling of fine sediments dynamics in a dam reservoir during a flushing event

L. Guertault, B. Camenen & A. Paquier
Irstea, UR HHLY, Centre de Lyon-Villeurbanne, Villeurbanne, France

C. Peteuil
CNR, Lyon, France

ABSTRACT: Regular flushes of the two Swiss Verbois and Chancy-Pougny dams on the Upper Rhône river are conducted since deposition in the reservoir endanger the city of Geneva. To mitigate Suspended Sediment Concentrations (SSC) peaks due to sediment release and limit sediment trapping, supporting operations are carried out along the river thanks to the developed French Génissiat dam. A 1D hydro-sedimentary model is applied to reproduce the fine sediments dynamics in the Génissiat reservoir during the 2012 flushing event. A stratification module is proposed to estimate the vertical concentration profile close to the dam and better reproduce SSC and fluxes at the dam outlets. Possible numerical improvements of the model are proposed to take into account processes related to fine sediment properties such as mud consolidation.

1   INTRODUCTION

Sediment trapping in reservoirs is a worldwide issue since it endangers hydropower installations. The low flow velocities in reservoirs reduce the transporting power of the stream and lead to large deposition of incoming sediment (Brown 1943; Morris and Fan 1998). Because of sediment trapping, one can observe a reduction of the reservoir capacity and downstream damages due to the sediment cut off. Dam flushing is a common technical solution used to reduce bed aggradation. During flushing operations, the water level of the reservoir is lowered to establish an accelerated flow allowing to erode and transport a part of the reservoir substrate (Di Silvio 2001). However, the release of a significant quantity of fine sediments downstream of the dam during flushes induces environmental damages. On the French Upper Rhône river, authorities set regulatory Suspended Sediment Concentrations (SSC) thresholds to prevent those damages, so that dam operators have to manage reservoirs to control SSC.

A better understanding of sediment transport dynamics is necessary to improve the reservoir management and control flushes outcomes. This study focuses on the Génissiat dam reservoir dynamics during flushes of a series of dams. In order to simulate hydro-sedimentary processes of this elongated reservoir, characterised by a deep and narrow channel, a 1D hydro-sedimentary model is applied. The main issue of this modelling is to reproduce SSC measured at the dam outlets located at different elevations using a module that reproduces the vertical stratification of SSC observed in the reservoir.

2   STUDY SITE AND THE 2012 FLUSHING EVENT

The Rhône river is a major river in Europe and flows in its upper part in Switzerland and then through France to Mediterranean sea. Downstream Lake Geneva, two dams are built on the Swiss Rhône river: the Verbois and Chancy-Pougny dams. These reservoirs trap a

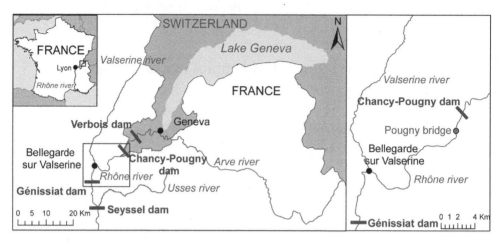

Figure 1. Location of the Upper Rhône river.

large quantity of sediment provided by the Arve River. The 70 m high Génissiat dam is located downstream on the French Upper Rhône River (Fig. 1).

Since the construction of the Verbois dam in the 40s, sediment flushing operations are regularly conducted on the Rhône River to prevent flood hazards in the lowest parts of Geneva city, as the reservoir bed aggradation leads to the rise of water levels (Peteuil et al. 2013). Downstream of the Swiss-French border, and more particularly at the Génissiat reservoir, supporting operations are carried out by the CNR to limit the impact of the high SSC released from the Swiss dams. The main challenge for CNR is to maintain an average concentration downstream of Génissiat below 5 g/L throughout the operation. Meeting the maximum concentration limits set by the French authorities is based on the dilution capacity of the Génissiat dam, thanks tothree hydraulic outlets located at different levels (a bottom gate, a half depth gate and a surface spillway). During flushing operations, an extensive measurement network is deployed by CNR on the French Upper Rhône River to monitor the operation progress. In particular, discharges and SSC chronicles are measured at Pougny and at the Génissiat dam outlets (Fig. 1).

Feedbacks from the numerous flushing operations have permitted to improve the management protocol of the Génissiat reservoir during flushes. The reservoir water level is first lowered to flush a volume of sediments and then raised again while Swiss reservoirs are flushed. Therefore, a part the sediments removed from the Swiss reservoirs are eventually trapped in the Génissiat reservoir. Suspended load and bedload sampling have shown that solid inflow and outflow during flushing events are mainly composed of fine sediments (clay, silt and fine sand). Those fine sediments have the largest contribution in the Génissiat reservoir volumetric budget (Guertault et al. 2014).

3 HYDRO-SEDIMENTARY MODEL

3.1 1D streamwise sediment transport model

The hydro-sedimentary model used in this study is Mage-AdisTS. The hydrodynamic module Mage solves 1D shallow water equations with Manning-Strickler formulation for roughness. An implicit numerical scheme is used to solve the equations, so that the model is not recommended for supercritical flows. The Adis-TS module solves advection-dispersion equations in conservative formulation (Equation 1) for several sediment grain sizes (Camenen et al. 2013). Those equations are coupled using source terms that allow to model erosion and deposition. The separation of main channel and medium zone dynamics is possible. Adis-TS is loosely coupled with Mage that provides the evolution of hydrodynamic parameters.

148

$$\frac{\partial SC_i}{\partial t} + \frac{\partial QC_i}{\partial x} - \frac{\partial}{\partial x}\left(D_f S \frac{\partial C_i}{\partial x}\right) = (E_i - D_i)W_z \tag{1}$$

where C_i is the concentration for grain size i, $C = \Sigma_i C_i$ is the total concentration, Q is the discharge, S is the wet section, D_f is the diffusion coefficient, E and D are erosion and deposition fluxes, W_z is the river width.

The source term combines (Partheniades 1965) formula for erosion and (Krone 1962) formula for deposition:

$$(E_i - D_i) = a_{PD,i}(C_{eq,i} - C)\frac{w_{s,i}}{h} \tag{2}$$

where $C_{eq,i}$ is the equilibrium concentration for the grain size i, $a_{PD,i}$ is a calibration parameter, $w_{s,i}$ is the settling velocity for the grain size i calculated using the Camenen formula (2007), h is the water depth.

The equilibrium concentration depends on the bed shear stress:

$$C_{eq,i} = \begin{cases} C_{0,i}\left(\dfrac{\tau}{\tau_{cr,i}} - 1\right) & \text{if } \tau/\tau_{cr,i} > 1 \\ 0 & \text{if } \tau/\tau_{cr,i} \leq 1 \end{cases} \tag{3}$$

where $C_{0,i}$ is a calibration parameter, τ is the bed shear stress, $\tau_{cr,i}$ is the critical shear stress for initiation if movement for the grain size i, estimated using (Soulsby and Whitehouse 1997) formula.

3.2 1D Vertical stratification model

For fine granulometric fractions such as clay and silt, the vertical profile of concentration is almost uniform, but for coarser sediment, the buoyancy force inhibits the vertical mixing by turbulence and leads to a vertical gradient in SSC (Wright and Parker 2004). Since the 1D transport model only estimates an average concentration for each cross section, a module is implemented to reproduce that vertical stratification and calculate the SSC at dam outlet positioned at different water depths.

The vertical concentration profile can be derived from the mass conservation equation and assuming a steady state vertical diffusion equation. Different formulations are obtained depending on the model used for the sediment diffusivity. Preliminary calculation have shown that the concentration at the dam half bottom gate can be estimated from the concentration at the dam bottom with an exponential profile:

$$C_i(z) = C_i(z_0)\exp\left[-\frac{w_{si}}{\epsilon_v}(z - z_0)\right] \tag{4}$$

where z_0 is the reference elevation, $\epsilon_v = \sigma_E \kappa u * h$ is the sediment diffusivity, σ_E is Schmidt number ($\sigma_E = 1$ as a first approximation), $\kappa = 0.41$ is Von Kàrmàn parameter, $u*$ is the shear velocity, h is thewater depth, $R_i = w_{si}/\kappa u *$ is Rouse number for grain size i.

Preliminary calculation have shown that the concentration passing through the dam half bottom gate can be estimated with an exponential profile applied to the concentration passing through the dam bottom gate.

Some assumptions were formulated to apply a stratification model to this study:

- section-averaged hydraulic parameters computed by the 1D model are assumed representative of the real local values,
- the average concentration passing through a gate can be approximated by the concentration calculated at the elevation of the center of gravity of the gate.

The stratification model implemented is described hereinafter. For each calculation time the shear velocity $u*$ is calculated at the outlets location using the results from the hydrodynamic model, and the Rouse number R_i is then calculated for each grain size. An average concentration in the cross-section for each grain size is provided thanks to the Adis-TS model. By definition, the average concentration is:

$$C_{m,i} = \frac{1}{h}\int_{z_0}^{z_0+h} C_i(z)dz \qquad (5)$$

Using Equations 4 and 5, an expression for the bottom concentration for each grain size i may be written such as:

$$C_i(z_0) = \frac{C_{m,i}\frac{R_i}{\sigma_E}}{1-\exp\left[\frac{R_i}{\sigma_E}\frac{(-h+z_0)}{h}\right]} \qquad (6)$$

Equation 4 is then applied at the outlet elevations to estimate the concentration for each grain size i in the different gates. For each gate, the total concentration is the sum of concentrations for all the grain sizes modelled.

4 2012 FLUSHING EVENT MODELLING

4.1 Calibration of the hydraulic model

The computational domain focused on the 24 km long reach from Pougny to the Génissiat dam (Fig. 1). The reservoir geometry was described with about 130 cross sections surveyed in December 2011. From that bathymetry, a 100 m regular mesh was built. Concerning the hydrodynamic model, upstream and downstream boundary conditions were respectively described by the flush hydrograph measured at Pougny and the water level measured at the Génissiat dam. Strickler coefficients are calibrated to adjust friction to fit waterlines measured during interflush periods and the 2012 flushing event (Fig. 2b). In upper gravel-bed reaches of the reservoir, the skin friction coefficient based on the grain size measured in the main channel with the Strickler formula (1923) seems to be representative of the coefficient used in the model (Fig. 2a). Friction is reduced in the downstream part of the reservoir. Approaching the dam, bed substrate is finer and the total Strickler coefficient is lower than the skin friction coefficient. The total coefficient takes into account other factors of flow resistance (such as bed forms or the river morphology), that become significant compared to the skin roughness.

4.2 Sediment input

Sediment diameters to be used in the model have been determined. During the 2012 flush, a few samples of suspended sediments have been collected at the water surface at Pougny and at the

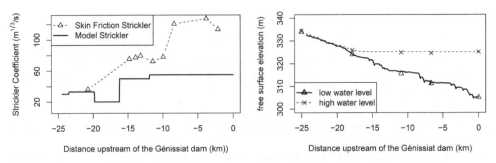

Figure 2. Calibrated Strickler coefficients (a) and computed waterlines for the 2012 flush (b).

Génissiat dam gates outputs (Lerch and Thizy 2013; Launay 2013). Assuming each granulometric mode can be represented by a normal distribution, a classification into 5 granulometric modes with median diameters from 4 µm to 400 µm allows to reproduce the total distribution of the samples. The upstream boundary condition has been described for each population. It is deduced from the total SSC measured at Pougny and weighted by the proportion of each population found in the samples. The reservoir bed substrate was described according to riverbed samples collected along the main channel of the reservoir before the flushing event (Fig. 3). A sublayer of fine sediment that could be eroded has been defined along the reservoir. Their thickness is the maximum thickness of the deposits in the subreach during the previous interflush period. Between 15 km and 10 km upstream of the dam, the sediment layer is composed by medium sand with a diameter of 400 µm (Fig. 3a). Closer to the dam, the sediment layer is a sediment mixture of fine sand, silt and clay. Thickness of the sediment layer increases approaching the dam (Fig. 3b).

Parameters related to the physical properties of the sediment were determined from field data or literature. Table 1 includes the properties of the sediment used in the model. Consolidation processes were not taken into account. A porosity was estimated for each grain size, however since the bottom layer consists of a mixture of several classes of sediment, a constant value $p = 0.45$ was considered for fine sediments.

4.3 Adis-TS model calibration

$C_{0,i}$ and $a_{PD,i}$ parameters have been calibrated for each grain size. An analysis of the asymptotic behaviour of the sediment transport law has been done:

- when $C_{eq,i} = 0$, $D_i = a_{PD,i} C w_{s,i}/H$, which means that $a_{PD,i}$ is the most significant parameter to calibrate deposition rate,
- when $C = 0$ and $\tau/\tau_{cr,i} \geq 1$, $E_i = a_{PD,i} C_{0,i} (\tau/\tau_{cr,i} - 1) w_{s,i}/H$, which means that both $a_{PD,i}$ and $C_{0,i}$ have an effect on erosion process calibration.

$a_{PD,i}$ was considered as a non-equilibrium adaptation coefficient as suggested by (Armanini and Di Silvio 1988) related to the fact that under unsteady flow conditions, the solid flow does not immediately reach the equilibrium, so that erosion and deposition mechanisms present an inertia. This coefficient represents the responsiveness of the evolution of the solid load compared with

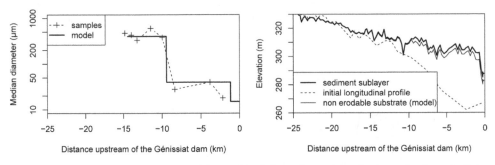

Figure 3. Median diameter (a) and thickness (b) of the sediment layer in the main channel of the reservoir in the model.

Table 1. Physical properties of the model sediment sizes.

Name	Clay	Fine silt	Coarse silt	Fine sand	Medium sand
d_{50} (m)	4	20	40	100	400
Settling velocity (m/s)	$9.0\ 10^{-6}$	$2.3\ 10^{-4}$	$9.0\ 10^{-4}$	$5.3\ 10^{-3}$	$3.6\ 10^{-2}$
Critical shear stress (Pa)	0.15	0.15	0.15	0.16	0.22
Porosity	0.45	0.45	0.45	0.45	0.4

the variations of the hydraulic conditions. $C_{0,i}$ was considered as a surface erosion rate constant, with the dimension of aconcentration, representative of the bottom sediment mobility.

According to surveys, the sediment load in the half depth gate is almost exclusively composed of clay and silt, when the sediment load in the bottom gate is also composed of sand. The first step is to calibrate $a_{PD,i}$ and $C_{0,i}$ for clay and silt for which the SSC vertical profile is homogeneous, to reproduce measured SSC and proportion of each fraction in the half depth gate. $a_{PD,i}$ and $C_{0,i}$ are then calibrated for sand to reproduce measured SSC and proportion of each fraction in the bottom gate.

4.4 Results

Calibration parameters used to model the 2012 flush are presented Table 2.

For fine sediments (d<0.1 mm), $a_{PD,i}$ decreases when grain size increases, until it reaches a constant value for sands. An interpretation for the quite high value used for clay is that this parameter counterbalances the settling velocity calculated for a single clay particle whereas flocculation may occur. $C_{0,i}$ also decreases when grain size increases.

From the average concentration (Fig. 4d) computed by the model Adis-TS, the stratification model estimates the concentration in the dam outlets. Erosion of the Génissiat reservoir during the first week of the flush is well reproduced in terms of SSC transport dynamics. It is not well reproduced in terms of values, as the model is not able to reproduce SSC peaks measured in the bottom gate. During the second week of the flush, when the Swiss reservoirs are drawn down, the model seems less efficient. The first SSC peak coming from Swiss reservoirs 8 days after the beginning of the operation is overestimated by the model. (Fig. 4a.b.c). An interpretation is that the 1D model is not able to reproduce the propagation and diffusion ofthe concentration that should delay and attenuate the concentration signal. Concentration is also underestimated in the bottom gate (Fig. 4a). It may be due to the fact that the upstream sediment inflow in the model is estimated from surfacesamples and may be unrepresentative of the real inflow because it underestimates a part of sand load which is not measured at the

Table 2. Calibration of C_0 and a_{PD} parameters.

Parameter	Clay	Fine silt	Coarse silt	Fine sand	Medium sand
$a_{PD,i}$	15	2	0.5	0.5	0.5
C_0,i	1.7	1	0.8	0.3	0.3

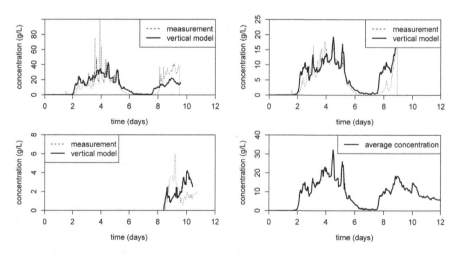

Figure 4. Measured and calculated concentration at Génissiat dam during the 2012 Flush at the: bottom gate (a), half depth gate (b), surface spillway (c) and average modelled concentration (d).

Table 3. Scores characterizing the calibration accuracy.

	Fluxes first phase (10^3 T)		Fluxes second phase (10^3 T)	
	Model	Measure	Model	Measure
Bottom gate	891	732 ± 220	648	421 ± 127
Half depth gate	477	466 ± 134	219	124 ± 37
Surface spillway	0	0	211	129 ± 39

surface. The signal at the spillway (Fig. 4c) is not well reproduced may be due to the fact that it is located on the right bank section average parameters are likely to represent it poorly.

Fluxes presented Table 3 illustrate former comments on Figure 4. The agreement between calculated and measured fluxes is good for the first phase, but the model overestimates the flux during the second phase for the three outlets.

5 DISCUSSION

The contribution of the stratification model is quite significant. It can be highlighted by comparing the total sediment mass passing through the dam outlets calculated with either SSC measured during the flush (1667 10^3 T), average SSC computed from Adis-TS model (4000 10^3 T), or SSC computed from the stratification model (2446 10^3 T). Even if the model overestimates fluxes at the Génissiat dam, the stratification model gives a better estimate of those fluxes than the only section averaged model. Several causes can explain differences between measured and calculated SSC. First, a finer sediment description of initial and boundary conditions thanks to additional data such as the quantification of the sand input at Pougny and a better reservoir bed substrate description could be useful. The calibration of parameters $a_{PD,i}$ and $C_{0,i}$ could also be improved.

Some improvement of the model may be done. For example, the calculation of local instead of section averaged shear stress could be implemented. Mud consolidation could be taken into account with the use of a different τ_{cr} for erosion for consolidated or fresh deposits. It may be worth to take into account fine particles aggregation because the settling velocity of flocs is higher than the one for single particles and should promote deposition. Several suggestions might help to reproduce SSC peaks, as for example the implementation of a bank failure module. Cores sampled close to the dam (Tissot and Merketa 1994) have shown that the bed substrate is composed of alternate layers of fine sediments and sands that can not be described by the model in which the bed substrate is homogeneous in the vertical. Occurrence of sandy layers should promote coarse sediment transport close to the dam and thus SSC peaks.

The 1D model may also be not accurate enough to reproduce correctly the main processes involved in fine sediments dynamics. Indeed, the presence of a large alluvial reach with a secondary channel at the upper part of the reservoir should lead to 2D hydro-sedimentary processes, inducing transversal diffusion, particularly during the second phase of the flush. Moreover, the configuration close to the dam is three dimensional, with a particular system of dam gates. The bottom gate is located 200 m upstream from the dam in the right side, the half depth gate is located 50 m upstream from the dam in the left bank and the surface spillway is located in the right bank of the dam. The 1D model is thus limited by its unidimensional description of hydraulic parameters, even if the stratification model takes into account the vertical dimension close to the dam.

6 CONCLUSION

A 1D hydro-sedimentary model is used to reproduce the fine sediment dynamics of the Génissiat reservoir during the 2012 flush. To address the dimensional limitation of the 1D model, especially in the vertical dimension in case of high water levels and low velocities close to the dam, a stratification model is proposed. The SSC vertical profile is calculated and allows to estimate the SSC at the dam outlets. The model accuracy is estimated by comparing the results to field measurements showing a significant contribution of the stratification model. Sediment fluxes at the dam are better reproduced.

As a validation, the model will be used to reproduce other flushes such as the 2003 flush. The sensibility to calibration parameters and the hydro-sedimentary description of the event for both the 1D and stratification model will also be evaluated. 1D model assumptions will be verified with 2D and 3D models close to the dam, particularly the currentology depending on the gates opening, and the shear stresses distribution within a cross section. Those results should help choosing further developments to improve the model among the solutions cited in the discussion.

ACKNOWLEDGEMENTS

Authors want to thank J.B. Faure from Irstea Lyon for developing the Mage Adis-TS model and field measurement teams of Irstea and CNR for collecting the data used to build the model.

REFERENCES

Armanini, A. & Di Silvio, G. 1988. A one dimensional model for the transport of a sediment mixture in non equilibrium conditions. *Journal of Hydraulic Research 26*: 275–292.
Brown, C., B. 1943. *The control of reservoir silting*. Washington, D.C. : U.S. Dept. of Agriculture.
Camenen, B. 2007. Simple and general formula for the settling velocity of particles. *Journal of Hydraulic Engineering 133*: 229–233.
Camenen, B., Andries, E., Faure, J., de Linares, M., Gandilhon, F. & Raccasi, G. 2013. Experimental and numerical study of long term sedimentation in a secondary channel: example of the Beurre island on the Rhône river, France. In *ISRS Kyoto, Japan*.
Di Silvio, G. 2001. Basic classification of reservoir according to relevant sedimentation processes. In *29th IAHR World Congress, Bejing, China*, pp. 285–293.
Guertault, L., Camenen, B., Peteuil, C. & Paquier, A. 2014. Long term evolution of a dam reservoir subjected to regular flushing events. *Advances in Geosciences 39*: 89–94.
Krone, R. 1962. Flume studies of the transport of sediment in estuarial shoaling processes: final report. Technical report, Hydraulic Eng. Lab and Sanitary Eng. Res. Lab., University of Califormia, Berkeley, California, USA.
Launay, M. 2013. *Fluxes of particulate contaminants in a large anthropized river: dynamics of PCB and mercury transported by suspended sediment load fin the Rhône river, from Lake Geneva to the Mediterranean sea [Flux de contaminants particulaires dans un grand cours d'eau anthropisé: dynamique des PCB et du mercure transportés par les matières en suspension du Rhône, du Léman á la Méditerranée] (in French)*. Ph. D. thesis, Claude Bernard University, Lyon 1.
Lerch, C. & Thizy, R. 2013. 2012 flush sediments analysis [Analyses des sédiments: Chasse 2012] (in French). Technical report, CNR.
Morris, G. & Fan, J. 1998. *Reservoir Sedimentation Handbook: Design and Management of Dams, Reservoirs and Watersheds for Sustainable Use*. McGraw-Hill.
Partheniades, E. 1965. Erosion and deposition of cohesive soils. *Journal of Hydraulic Division 91*: 105–139.
Peteuil, C., Fruchart, F., Abadie, F., Reynaud, S., Camenen, B. & Guertault, L. 2013. Sustainable management of sediment fluxes in reservoir by environmental friendly flushing: the case study of the Genissiat dam on the Upper Rhône river (France). In *ISRS Kyoto, Japan*, pp. 1147–1156 (CDRom).
Soulsby, R. & Whitehouse, R. 1997. Threshold of sediment motion in coastal environments. In *Pacific Coasts and Ports: Proceedings of the 13th Australasian Coastal and Ocean Engineering Conference and the 6th Australasian Port and Harbour Conference*, Volume 1, pp. 145–150. Centre for Advanced Engineering, University of Canterbury, Christchurch, N.Z.
Strickler, A. 1923. Contributions to the question of a velocity formula and roughness data for flumes, channels and closed pipelines [Beitrhe zur frage der geschwindigkeitsformel und der rauhigkeitszahlen fur strme, kanle and geschlossene leitugen] (in German). In *Releases of the Federal Office for Water Management*.
Tissot, J., P. & Merketa, B. 1994. Géotechnical analysis of cores sampled in the Genissiat reservoir [Étude géotechnique de sondages dans la retenue de Génissiat] (in French). Technical report, ENSG, Geomechanics Laboratory.
Wright, S. & Parker, G. 2004. Flow resistance and suspended load in sand-bed rivers: Simplified stratification model. *Journal of Hydraulic Engineering 130*: 796–805.

Reservoir Sedimentation – Schleiss et al. (Eds)
© *2014 Taylor & Francis Group, London, ISBN 978-1-138-02675-9*

Numerical analysis of sediment transport processes during a flushing event of an Alpine reservoir

G. Harb, C. Dorfmann & J. Schneider
Graz University of Technology, Graz, Austria

H. Badura
VERBUND Hydro Power AG, Vienna, Austria

ABSTRACT: This study focuses on the sediment transport processes during a flushing event of an Alpine reservoir. The reservoir had an initial storage volume of about 1.4 Mio. m³. However, echo-soundings performed in 2007 showed that approximately 890,000 m³ of sediments are already deposited in the reservoir. This represents an annual sedimentation rate of about 6.1 percent of the initial reservoir volume. An open source three-dimensional numerical model was used to simulate the flushing process. Echo-soundings performed before and after the flushing event were used to set up the simulation and validate the morphological bed changes calculated by the numerical model. Additionally, an extensive sensitivity analysis was carried out by testing several sediment transport formulae. It was found that the simulations using the bed load transport formulae derived by Van Rijn and Meyer-Peter Müller showed the best agreement with the measured data.

1 INTRODUCTION

Reservoir sedimentation is a problem in many Alpine reservoirs. In the case of Alpine reservoirs with a small storage volume compared to the annual inflow, such as reservoirs of run-of river power plants, the water depth are usually smaller compared to reservoirs of storage and pump-storage hydro power plants. A larger part of the suspended sediments is thus transported through the reservoir and depositions of bed load fractions represent the main problem. The deposition of coarse sediments at the head of the reservoir may cause problems regarding flood protection by raising the bed level and thus, raising the water level too.

Flushing is one of the most common ways to manage sediment depositions in reservoirs. In most cases the flushing of reservoirs in the Alpine area is a special challenge because of the massive coarsening of the depositions from the weir to the head of the reservoir and the subsequently wide grain size distribution. At the head of the reservoir the gravel fractions are deposited, whereas silt and clay particles deposit in front of the weir.

The application of numerical models became state of the art for water flow calculations in river engineering in the meantime. However, the modelling of sediment transport processes still needs further development. In previous studies two-dimensional numerical model were used for the simulation of flushing process of the Bodendorf (Badura, 2007), Kali Gandaki reservoir in Nepal (Olsen, 1999) and the Leoben reservoir in Austria (Harb et al. 2012). Due to the increasing computer power in the last years three-dimensional models are used for these applications. Examples are the simulation of the sediment transport in the Three Gorges project (Fang and Rodi, 2003), the simulation of the sediment transport in the Feistritz reservoir (Dorfmann & Knoblauch, 2009) and the simulation of reservoir flushing in the Angustura reservoir (Haun & Olsen, 2012; Haun et al. 2013).

An open source three-dimensional numerical model, coupled with a morphological module was used in this study to simulate the flushing process in the reservoir of the HPP Fisching in

Austria. The calculated morphological bed changes are validated with measurements, taken by sonar, and presented in this study.

2 PROJECT AREA AND BACKGROUND

2.1 *Description of the reservoir*

The reservoir Fisching is approximately 4.5 km long with an initial storage volume of about 1.4 Mio. m³ in 1994. In the last years, the mean annual sediment deposits are approximately 85,000 m³. A small amount of the deposited sediments has been eroded and transported through the reservoir during former flushing events in the years 1999 and 2002, and a partial flushing in the year 2004. However, echo-soundings performed in the year 2007 showed that 890,000 m³ of sediments are already deposited in the reservoir. This represents an annual sedimentation rate of about 6.1 percent of the initial reservoir volume.

Figure 1 shows an overview of the meandering reservoir with the sediment sampling points (see Chapter 2.2). The river Mur at Fisching has an mean annual discharge of 48.2 m³/s and the 1-year flood has a discharge of 240 m³/s.

2.2 *Sediment sampling*

Seven representative sediment samples were taken from the reservoir. The freeze-core method was used for taking the sediment samples, starting from the weir (P1, sample 1) up to the head of the reservoir (P7, sample 7).

Figure 1.   Reservoir with sediment sampling points P1–P7 in plan view.

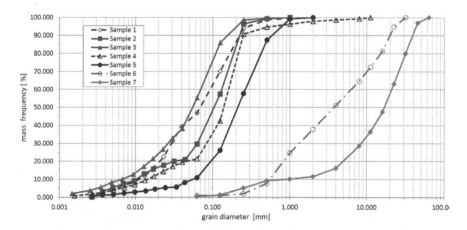

Figure 2.   Grain size distributions of the sediment samples taken from the reservoir.

The samples show the effect of fractionized sedimentation and thus, large variations in the grain sizes between Sample 1 and Sample 7 (Fig. 2). The d_m of Sample 1 is below 0.1 mm and the fractions > 1 mm are 0%. Whereas the d_m of Sample 7 is about 18 mm and the fractions > 1 mm are 90%. These large variations in the grain sizes in the reservoir are one of the major challenges in the numerical sediment transport modeling.

3 NUMERICAL SIMULATIONS

3.1 Numerical model

The numerical simulations were performed with TELEMAC-3D, which solves the three-dimensional incompressible Navier-Stokes equations for free surface flow. A detailed description of the theoretical aspects used in TELEMAC is given in Hervouet (2007). TELEMAC-3D extends the two-dimensional triangular mesh to the vertical dimension by the implementation of a number of vertical planes or levels. Several turbulence models are implemented in TELEMAC-3D. In this study for turbulence closure the standard k-ε model was used. The Strickler friction law was applied to compute the energy losses caused by bottom friction with a constant Strickler value of 35 $m^{1/3}$/s for the whole computational domain. The simulations were performed with a semi-implicit time integration with a time step of 1.0 seconds, which resulted in maximum Courant numbers of approximately 3.

The morphological module SISYPHE was internally coupled with TELEMAC-3D to compute the sediment transport. The evaluation of the sediment transport functions compared with the measured erosion and deposition pattern in the reservoir is presented later.

3.2 Input data

Based on the existing bathymetry data, a three-dimensional digital elevation model was generated. The triangular mesh with approximately 46,300 cells and an average edge length of 4 m was generated with the free software BlueKenue (CHC 2010). The bottom heights of the digital elevation model were linearly interpolated onto the mesh nodes. The partial flushing event monitored in 2009 (nearly 1-year flood) was simulated in this study. The water level at the weir was lowered max. 1,6 m for 37 hours. In the simulation of the flushing event 48 hours were computed. Both water level and the inflow hydrograph are shown in Figure 3. The computation time was between 58 hours on a small laptop and 64 minutes on 32 nodes of a cluster (VSC2).

The sediment conditions in the numerical model were based on the sediment distribution in the reservoir. In total, 8 sediment sizes with spatial varying fractions were used as

Figure 3.   Freece-core samples P1 and P7.

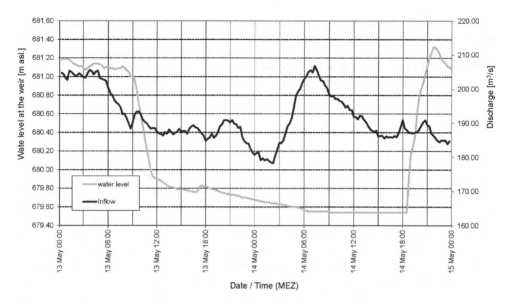

Figure 4. Inflow to the reservoir (right) and corresponding water level at the weir (left).

initial condition for the calculation of the sediment transport. The thickness of the active sediment layer was chosen with 10 cm according to the larger fractions of the sediment. A sensitivity analysis with thicknesses of 200, 100, 50 and 10 cm showed no significant changes of the results.

3.3 Calibration of the numerical model

The calibration of the hydrodynamic model was done using ACDP measurements performed at the prototype to calibrate the roughness at the river bed and at the banks (Harb et al. 2013).

4 RESULTS AND DISCUSSION

4.1 Evaluation and analysis of the measured bed changes

Echo-soundings performed before and after the flushing event were used to set up the numerical model and validate the morphological bed changes calculated by the model.

The measured bed levels before (Fig. 5) and after the flushing event (Fig. 7) were used to calculate the changes in the bed levels and the erosion pattern. The differences in the bed levels (Fig. 6) were used for validation of the numerical model. These measurements are very interesting, because at the flushing event 2009 most of the erosion occurred at the inner site of the river bends and not on the outer side as expected. This effect may be caused by the complex meandering geometry of the reservoir and the parameters of the flushing operation itself. The flushing operation 2009 was performed with a maximum discharge of about 200 m³/s (nearly 1-year flood) and a maximum lowering of the water level at the weir of only 1.1 m. The water level in the reservoir was thus relatively high and the bed shear stress at the outer site of the bends was apparently not sufficient enough to initiate erosion in this area and to deepen the flushing channel.

The massive depositions at the inner site of the river bend were eroded due to the higher discharge. Sand slides and instabilities at the river banks lead to erosion on the inner site and deposition on the outer site of the river bend.

158

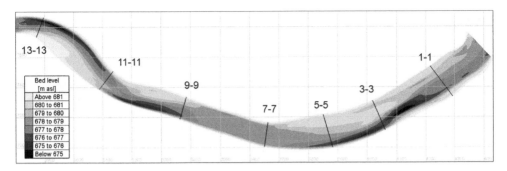

Figure 5. Measured bed levels before the flushing event 2009.

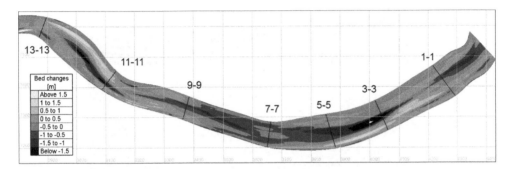

Figure 6. Measured bed changes after the flushing event 2009.

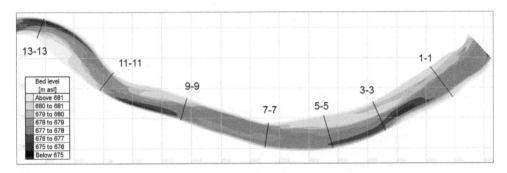

Figure 7. Measured bed levels after the flushing event 2009.

4.2 *Numerical results of the flushing simulation*

For the modelling of the sediment transport a sensitivity analysis was carried out. The following parameters were tested and varied:

- Sediment transport formulae (Meyer-Peter-Müller, Engelund-Hansen and Van Rijn)
- Skin friction correction, which takes the effective grain shear stress into account
- Slope effect formula and deviation formula, which takes the gravity of the sediment grain on lateral slopes into account

The sensitivity analysis showed that the skin friction correction and the deviation formula affect the results significantly. Without the deviation formula the erosion pattern is inverted due to the secondary currents effect and the direction of the shear stress vectors. Figure 8 shows the calculated bed changes using the sediment transport formula of Meyer-Peter-Müller.

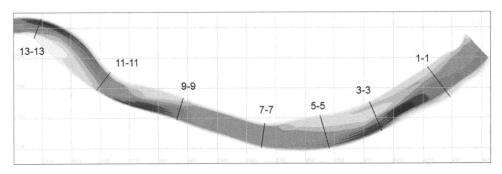

Figure 8. Calculated bed levels using the sediment transport formula of Meyer-Peter-Müller.

Figure 9. Observed standing waves upstream of the weir during the flushing event 2012.

The comparison with the measured bed changes shows a good agreement. However, in all simulations so called "furrows" appeared in the streamwise direction (see Figs. 8 and 10). The reason for this numerical effect lies in the implementation of the Navier-Stokes equations and the boundary conditions, which lead to very small differences in the water level. This small differences cause differences in the bed shear stresses. Integrated over the total simulation time of 48h this effect results in different erosion rates of neighbour nodes in the domain.

Figure 10 shows the measured and the calculated bed changes in different cross sections in the reservoir (locations see Fig. 8). The calculated bed changes in sections 1–1 and 11–11 obtained by the Meyer-Peter-Müller formula (1948) and the formula of van Rijn (1984a) show very good agreement with the measured bed changes. Nevertheless the calculated bed levels tend to be too high in other sections, which reflect too low erosion rates. This difference was much higher in the first simulations and is caused by the implemented skin friction algorithm, which reduces the grain shear stress based on the grain size of the sediment. This skin friction correction does not take the increased roughness due to occurring bed forms into account. Therefore, the calculated sediment transport rates with the implemented skin friction correction algorithm are quite low in the areas with very fine sediments, which is a result of the low effective bed shear stress (amounts to approximately 20% of the total shear stress).

Due to the first results the skin friction algorithm was adapted to improve the results compared to older calculations (Harb, 2013). However, it was not able to take the whole complexity of the bed forms and their effect on the bed roughness and sediment transport during flushing events into account. Observations during a flushing event in 2012 showed for example unstable standing waves, which indicate antidunes, upstream of the weir (Fig. 9). The calculated bed changes were thus too low in some sections.

Section 1-1 – river bend in front of the weir with outer bank on the right side

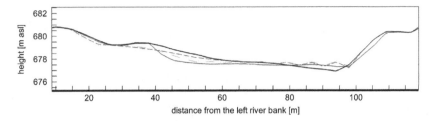

Section 3-3 – river bend with shallow water zone on the left bank and outer bank of the right side

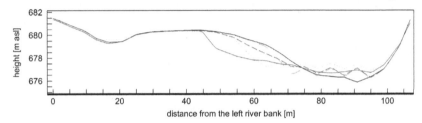

Section 9-9 – straight river section

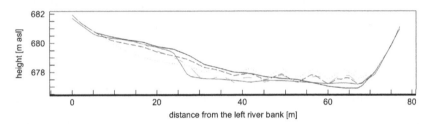

Section 11-11 – river bend with outer bank on the left side

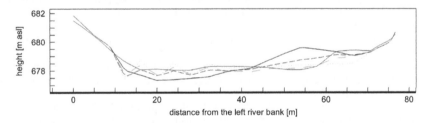

Figure 10.   Cross sections of the measured bed levels before (dark grey line) and after (light grey line) the flushing event 2009 compared with the calculated bed levels using the formulae of Van Rijn (long dashed), Engelund-Hansen (dotted) and Meyer-Peter-Müller (short dashed); location of the sections can be found in Figures 5–8.

Simulations without skin friction correction, that means using the total shear stress as input for SISYPHE, resulted in too large erosion patterns. A variation of different hiding and exposure formulae showed only marginal differences. The calculated bed changes using Engelund-Hansen total load transport formula (1967) showed the highest deviations to the measured bed changes thus the results are not shown in Figure 10. Although the three-dimensional model is able to take the secondary currents into account, the measured erosion pattern could not be reproduced in case of the Engelund-Hansen total load formula. The use of a total load formula may not reflect the different sediment transport behavior of the bed load and the suspended load fractions and the intense three-dimensional effects in this case.

5 CONCLUSIONS

This paper discusses the simulation of a flushing event in an Alpine reservoir. The open source numerical model TELEMAC-3D combined with the morphological module SISYPHE was used to simulate the flushing process. The results of the numerical model were compared with the measured bed changes. The results of the measurements in the reservoir showed that erosion occurred at the inner site of the river bends in this case. Combined effects of a complex meandering geometry of the reservoir, partial drawdown water level, sand slides and instabilities at the river banks supposed to be the reasons for this observed effect.

The sediment transport formulae of Engelund-Hansen, Meyer-Peter-Müller and Van Rijn were used to model the sediment transport. The calculated bed changes derived by Meyer-Peter-Müller and Van Rijn formula showed the best agreement with the measured bed changes. The use of a total load formula like Engelund-Hansen cannot be recommended in this case.

However, the measured erosion on the inner side and deposition on the outer side of the river bends in the meandering reservoir could be reproduced with the numerical model, but the calculated erosion rates tend to be too low. This effect is caused by the variation of the effective grain shear stress depending on the bed roughness and the grain sizes, which are changing during the flushing process. The improvement of the skin friction algorithm to take the complexity of the occurring bed forms during reservoir flushing into account could improve the result further. However, measurements of the sediment transport rates of the bed forms during the flushing event are not available and are difficult to perform, but would enhance the further understanding and the simulation of the sediment transport processes during flushing events.

REFERENCES

Badura, H. 2007. *Feststofftransportprozesse während Spülungen von Flussstauräumen am Beispiel der oberen Mur,* Dissertation, Graz University of Technology, Austria.

CHC—Canadian Hydraulics Centre, National Research Council, 2010. *Blue Kenue, Reference Manual,* August 2010.

Copeland, R.R. & Thomas, W.A. 1989. *Corte Madera Creek Sedimentation Study. Numerical Model Investigation.* US Army Engineers Waterways Experiment Station, Vickings-burg, MS.

Dorfmann, C. & Knoblauch, H. 2009. A Concept for Desilting a Reservoir Using Numerical and Physical Models. Water Engineering for a Sustainable Environment, *Proceedings of the 33.IAHR Congress,* Vancouver, Canada.

Engelund, F. & Hansen, E. 1967. *A Monograph on Sediment Transport in Alluvial Stream,* 1–63. Teknisk Forlag, Copenhagen V, Denmark.

Fang, H-W. & Rodi, W. 2003. Three-dimensional calculations of flow and suspended sediment transport in the neighborhood of the dam for the Three Gorges Project (TGP) reservoir in the Yangtze River. *J. Hydraulic Res.* 41(4), 379–394.

Harb, G., Dorfmann, C., Badura, H., & Schneider, J. 2013. Numerical Analysis of the Flushing Efficiency of an Alpine Reservoir. *Proceedings of the of 2013 IAHR World Congress, Chengdu, China.*

Harb, G., Dorfman, C., Schneider, J., Haun, S., & Badura, H., 2012. Numerical analysis of sediment transport processes in a reservoir. *In Proceedings of the River Flow 2012, San Jose, Costa Rica.*

Harb, G., 2013. *Numerical Modeling of Sediment Transport Processes in Alpine Reservoirs.* PhD Thesis. Graz University of Technology.

Haun, S. & Olsen, N.R.B. 2012. Three-dimensional numerical modelling of reservoir flushing in a prototype scale. *International Journal of River Basin Management.* 10(4) 341–349.

Haun S., Kjærås H., Løvfall S., & Olsen N.R.B. 2013. Three-dimensional measurements and numerical modelling of suspended sediments in a hydropower reservoir. *Journal of Hydrology,* Vol. 479, pp. 180–188.

Hervouet J-M. 2007. *Hydrodynamics of free surface flows: modelling with the finite element method.* Wiley.

Meyer-Peter, E. & Müller, R. 1948. *Formulas for bed-load transport. Proceedings of the 2nd Meeting of the International Association for Hydraulic Structures Research.* pp. 39–64.

Olsen, N.R.B. (1999a). Two-dimensional numerical modelling of flushing processes in water reservoirs. *J. Hydraulic Res.* 37(1), 3–16.

Van Rijn, L.C. 1984a. Sediment Transport. Part I: Bed load transport. *Journal of Hydraulic Engineering,* 110(10), pp. 1431–1456.

Reservoir Sedimentation – Schleiss et al. (Eds)
© *2014 Taylor & Francis Group, London, ISBN 978-1-138-02675-9*

Modelling suspended sediment wave dynamics of reservoir flushing

T.H. Tarekegn
Department of Civil, Environmental and Mechanical Engineering, University of Trento, Trento, Italy
School of Geography, Queen Mary, University of London, London, UK

M. Toffolon & M. Righetti
Department of Civil, Environmental and Mechanical Engineering, University of Trento, Trento, Italy

A. Siviglia
Laboratory of Hydraulics, Hydrology and Glaciology (VAW), ETH Zurich, Zurich, Switzerland

ABSTRACT: Flushing of fine sediments from reservoirs is one of the most effective techniques to reduce reservoir sedimentation, but the sudden release of high suspended sediment concentration can have adverse effects on the receiving water body. Field observations of sediment flushing operations have shown that the released volume of water and sediments propagate downstream through different types of waves. In particular the hydrodynamic signal travels faster than the sediment signal resulting in the splitting of the two waves. Since the sediment wave lags behind, deposition is enhanced in the tail of the hydrodynamic wave, where velocity decreases and cannot sustain the sediment in suspension: the separation phase of the two waves controls the deposition process of the released suspended sediments. The separation and the interaction between the two waves, especially in the transport of suspended sediments, is controlled by the sediment wave celerity. Neverthless in many sediment transport models the sediment wave celerity is assumed to be the mean flow velocity. We developed a simplified one-dimenional numerical model to study the interaction between the two waves. In the model we introduced a celerity factor that corrects the depth-averaged sediment transport velocity as a function of the shape of the vertical velocity profile and suspended sediment concentration. We observed that the use of the celerity factor in one-dimensional sediment concentration transport enhances the deposition because of the reduced celerity of the sediment wave, which separates sooner from the hydrodynamic wave.

1 INTRODUCTION

Reservoir sedimentation is one the most challenging problem in reservoir management with annual worldwide loss rate of reservoirs storage capacity by 0.5–1.9% (Liu et al. 2004). To overcome this problem various active and passive techniques have been developed such as flushing, sluicing, dredging and soil conservation and watershed management. Among these, sediment flushing is the most effective, economic and fast method to reduce reservoir sedimentation (Liu et al. 2004). However it may often have significant effects on downstream river reaches due to the high flows induced by the sudden release of water and high concentration of fine sediments (Espa et al. 2013). The most common downstream impacts include alteration of channel morphology (Brandt 2005), destruction of habitat type and stability (Wohl & Rathburn 2003), and river bed clogging (Liu et al. 2004). Generally adverse impacts of flushing releases depend mainly on the level of suspended sediment concentration and the modification of river habitats following sedimentation of the flushed material (Espa et al. 2013). Thus reservoir sediment flushing operation should be designed so as to safeguard the riverine ecosystems and at the same time reduce loss of storage of reservoirs. Designing such successful flushing operation needs better understanding of how the released sediments are transported and interact with downstream riverine ecosystems. The interaction between the hydrodynamic and sediment waves released from the dam plays important role in the deposition of the suspended sediments (Brandt 2005). Hence, characterising the dynamics of the two waves is expected to improve our understanding of the process and also to identify better flushing scenario. The main focus of this study is to model the wave dynamics and sediment deposition pattern and evaluate the effect of sediment wave celerity on the wave dynamics and deposition.

We first describe the main processes downstream of the dam during flushing operation in Section 2. The overview of the numerical model formulation and its components are presented in Section 3. In Section 4, detailed description of wave celerity derivation is presented. Finally, in Section 5 the effect of sediment wave celerity on wave dynamics is discussed.

2   DOWNSTREAM PROCESSES OF RESERVOIR FLUSHING

During sediment flushing operations two pulses are typically released in the downstream reach: a larger liquid discharge and a considerable amount of solids. These two pulses can be seen as waves travelling with their own celerities. The first wave transports the signal of a hydrodynamic variable (depth, discharge or velocity) with the celerity of flood waves, thus higher than the cross-sectional averaged flow velocity (e.g. Toffolon et al. 2010). The second wave is associated with the concentration of suspended sediments or dissolved substances, water temperature, etc. Since its celerity is equal, or smaller as in the case of suspended sediments than the flow velocity, the two waves travel with different speeds and tend to separate. For instance, this behaviour is noticeable downstream of hydropeaking releases characterized by different temperatures (Toffolon et al. 2010, Zolezzi et al. 2011). Field investigation of releases during flushing operations downstream of the Cachi reservoir in Costa Rica similarly show that there is a significant time lag between the suspended solid and liquid discharge where the liquid phase travels faster than the solid phase (Brandt & Swenning 1999, Brandt 2005). The phase lag increases as the two waves travel downstream. For instance, suspended-sediment concentration peaks lagged liquid peaks by about 2, 5 and 7 hours at 10, 30 and 70 km respectively downstream of the Cachi reservoir. We depict the typical dynamics of discharge Q and sediment concentration C in Figure 1. Figure 1b shows the conceptual representation of the longitudinal behaviour of the hydrodynamic and suspended sediment waves that propagate downstream of the dam. The temporal evolution (Fig. 1a) at downstream stations reflects in the spatial lag between the two waves (Fig. 1b), which grows while they propagate along the river channel downstream of a dam.

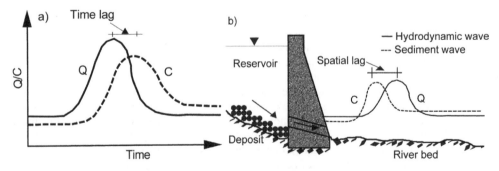

Figure 1.   Conceptual sketch of at-station discharge Q and sediment concentration C (a) and longitudinal profiles of suspended sediment and hydrodynamic waves of sediment flushing (b).

In Figure 1a the sediment concentration C does not increase as the discharge Q increases which is often observed in fluvial sediment transport. In the case of sediment transport by sediment flushing sediment peak occurs on the falling limb of Q. This phenomena is also noted in field observations (e.g. Brandt & Swenning 1999). The large amount of sediment concentration C in the recession limb of the discharge Q results deposition of sediments.

The time lag in the two waves and their interaction control the deposition of the suspended sediments carried by the release. In particular the hydrodynamic signal travels faster than the sediment signal resulting in the splitting of the two waves. Since the sediment wave lags behind, deposition is enhanced in the tail of the hydrodynamic wave, where velocity decreases and cannot sustain the sediment in suspension: the separation phase of the two waves controls the deposition process of the released fine sediments. Nevertheless, the mechanism how the temporal and spatial pattern of the deposition is controlled by the interaction between the two waves is not yet clear. On one hand, previous studies (Rathburn & Wohl 2001, Liu et al. 2004) focused mainly on the study of morphological evolution and suspended sediment concentration after sediment flushing. On the other hand, in many models the celerity of sediment waves is assumed equal to the depth averaged flow velocity, but it is important to recognize that this assumption might be misleading especially in modelling suspended sediment transport since sediments in suspension are transported mainly near the bed. We emphasize that proper description of the sediment wave celerity is required as it controls the interaction between the two waves. Thus the sediment wave celerity needs to be corrected based on a factor which depends on the shape of the concentration and velocity profile. Thus, we developed a one-dimensional (1-D) suspended sediment transport model that incorporates the correction factor. The model is applied to study the interaction between the two waves and the effect of the correction factor on the wave dynamics. This factor, called 'celerity factor' hereafter, is calculated analytically assuming an exponential concentration profile. In this work, we assumed a unerodible bed on which incoming suspended sediments deposit and re-suspend from this bed depending on the flow hydrodynamics. It is expected that characterizing and analysing the wave dynamics will help to understand deposition processes and pattern due to sediment flushing and thus to plan proper flushing operations that reduce negative impacts.

3 MODEL FORMULATION AND NUMERICAL SOLUTION

3.1 *Model description*

We formulated the model for a simple case as we are interested in the large scale dynamics of sediment transport. We introduced a 1D description of sediment transport during flushing operation where a given amount of water and fine sediments are released in a limited period of time. We assume the modelled river as wide rectangular channel. The hydrodynamic wave is described by 1D Saint-Venant shallow water equations (Equations 1–2) and the transport of suspended sediments by the continuity equation of sediments (Equation 3). The advection-reaction part of the equations is solved by weight average flux (WAF) method (Toro 1989) and the diffusion part by Crank-Nicolson method where the two are coupled by an operator splitting technique (Siviglia & Toro 2009).

$$\frac{\partial A}{\partial t} + \frac{\partial Q}{\partial x} = 0 \tag{1}$$

$$\frac{\partial Q}{\partial t} + \frac{\partial}{\partial x}\left(\beta\frac{Q^2}{A}\right) + gA\frac{\partial H}{\partial x} - gA\left(S_0 - S_f\right) = 0 \tag{2}$$

$$\frac{\partial}{\partial t}(CA) + \frac{\partial}{\partial x}(\alpha CQ) = \frac{\partial}{\partial x}\left(KA\frac{\partial C}{\partial x}\right) + B(E - D) \tag{3}$$

where t = time; x = longitudinal spatial co-ordinate; β = coefficient accounting for the deviation of local values of momentum from its cross-sectional average; g = gravitational acceleration; S_0 = local bottom slope; S_f = local energy slope; H = averaged depth of flow; C = depth-averaged volumetric sediment concentration; α = correction factor for sediment wave celerity; K = longitudinal diffusion/dispersion coefficient; B = channel top width; E = sediment entrainment rate; and D = sediment deposition rate.

The sediment exchange $E - D$ near the bed is evaluated at the conventional level z_r where the near bed suspended sediment concentration is estimated. The net flux of sediment across such an interface can produce a variation of the bed elevation ζ

$$\frac{\partial \zeta}{\partial t} = \frac{(D - E)}{1 - p} \tag{4}$$

where ζ = the bed elevation; and p = bed porosity.

3.2 Model closures

In order to close the governing equations of the model closure relations that are used to calculate the variables S_f, E, and D as a function of unknown variables A, Q and C are required. The diffusion coefficient is estimated by Fischer et al. (1979) formula for longitudinal dispersion. The friction slope S_f is calculated by Chézy equation

$$S_f = \frac{Q^2}{g A^2 \chi^2 R} \tag{5}$$

where R = hydraulic radius; and χ = Chézy coefficient. The bed resistance of the sand deposited bed is computed in the form of Chézy coefficient by using Wright & Parker (2004) formulations for the specified discharge, sediment grain size and bed slope. A constant roughness coefficient (Table 1) is used in case of no deposition. The following empirical relationships are used to estimate D and E (Celik & Rodi 1988).

$$D = w C_b; \quad E = w C_E \tag{6}$$

where w = effective settling velocity of a single particle (Richardson & Zaki 1954); C_b = near bed concentration; and C_E = equilibrium near-bed concentration. C_E is estimated by Van Rijn (1984) formula which is widely used in large scale applications (e.g. Duan & Nanda 2006).

Table 1. Input datasets.

Model parameters	Value
Base unit discharge q_0	0.8 $m^2 s^{-1}$
Peak unit discharge q_p	4.8 $m^2 s^{-1}$
Release duration t_p	3 hr
Bed slope S_0	0.007
Base sediment concentration C_0	0 $g l^{-1}$
Peak sediment concentration C_p	300 $g l^{-1}$
Median diameter of the fine sediments d_{50}	230 μm
Bed porosity p	0.4
Manning roughness of the unerodible bed n	0.021 $m^{-1/3} s$
Simulation period	55 hrs
Spatial steps Δx	100 m

4 SEDIMENT WAVE CELERITY

Many suspended sediment transport models assume that sediment advection occurs with the depth-averaged flow velocity $U = Q/A$ (e.g. Cao & Carling 2002). However, since the vertical distribution of both longitudinal velocity and suspended sediment concentration is not uniform, the depth-integrated transport is

$$Q_s = \int_{z_r}^{H} cu\,dz = \alpha CUH \tag{7}$$

where $u(z)$ = vertical profiles of velocity; $c(z)$ = vertical profiles of concentration; z = vertical coordinate with origin at the bed level (ζ); z_r = the reference level where the near bed suspended sediment concentration is estimated; H = averaged depth of flow; and α = celerity factor already introduced in Equation 3.

Since velocity increases upward approximately following a logarithmic profile while the suspended sediment concentration is maximum toward the bottom, the overall transport is less that the product CUH that could be expected for a tracer uniformly distributed along the vertical. Thus it is important to estimate the value of α

$$\alpha = H\frac{\int_{z_r}^{H} cu\,dz}{\int_{z_r}^{H} u\,dz\int_{z_r}^{H} c\,dz} \tag{8}$$

which contains the Einstein integrals (Einstein 1950).

In order to express the sediment wave celerity explicitly, Equation 3 can be rewritten as

$$\frac{\partial C}{\partial t} + \alpha U\frac{\partial C}{\partial x} = \frac{1}{A}\frac{\partial}{\partial x}\left(KA\frac{\partial C}{\partial x}\right) + \frac{(E-D)}{H} - (1-\alpha)\frac{C}{A}\frac{\partial A}{\partial t} \tag{9}$$

Equation 9 shows that the sediment wave celerity is αU, which is in general slower than the depth-averaged velocity U; hence, the separation between the hydrodynamic wave and the sediment wave is enhanced. An additional term arises in Equation 9 when $\alpha \neq 1$. This term can increase the concentration in the falling limb of the hydrodynamic wave producing values higher than the previous maximum despite no net erosion from the bed, an occurrence that is usually not possible in standard advection-diffusion problems.

In general, the value of α from Equation 8 cannot be calculated analytically. However, an approximate solution can be obtained by assuming logarithmic profile of velocity

$$u = \frac{u_*}{k}\ln\left(\frac{z}{z_0}\right) \tag{10}$$

where $z_0 = 0.033k_s$; u_* = bed-shear velocity; k = the von Karman constant; and k_s = bed roughness. We assumed that the bed roughness to be equal to grain roughness. Thus we adopted $k_s = 3d_{90}$ (Van Rijn 1984) where $d_{90} = 90\%$ particle diameter of sediment material. The simplified Rouse profile for the suspended sediment concentration

$$c = C_b\left(\frac{z}{z_r}\right)^{-Ro} \tag{11}$$

based on a linear diffusivity profile, where $Ro = w/ku_*$ is termed Rouse number, where w = particle settling velocity. The reference level (z_r) is also taken as $k_s = 3d_{90}$. Huybrechts et al. (2010) propose a simplification of the Einstein integrals that provides the analytical solution as

$$\alpha = -\left(\frac{I_2 - I_1 \ln\left(\dfrac{z_r^*}{30}\right)}{I_1 \ln\left(\dfrac{ez_r^*}{30}\right)} \right) \tag{12}$$

where $z_r^* = \dfrac{z_r}{h}$ and the two integrals read

$$I_1 = \begin{cases} -\ln z_r^* & \text{for } Ro = 1 \\ \dfrac{1 - z_r^{*(1-Ro)}}{1 - Ro} & \text{for } Ro \neq 1 \end{cases}, \qquad I_2 = \begin{cases} -\left(\ln z_r^*\right)^2 / 2 & \text{for } Ro = 1 \\ \dfrac{I_1 - z_r^{*(1-Ro)} \ln z_r^*}{Ro - 1} & \text{for } Ro \neq 1 \end{cases} \tag{13}$$

5 RESULTS AND DISCUSSIONS

5.1 Inputs and initial conditions

We adopt most of model input datasets from the Cachí Reservoir flushing observations in Costa Rica (Brandt & Swenning 1999) despite the lack of detailed data on flushing. The model inputs are summarized in Table 1. The solid and liquid discharges released at the upstream boundary are assumed to be a rectangular shaped hydrograph and sedimentograph. We define the initial concentration as clean water ($C = 0$) with uniform flow depth along the stretch of the river, whose slope is assumed constant. The river channel geometry is assumed to be a wide rectangular channel thus we consider discharge per unit width ($q = Q/B$). A simulation domain length of 500 km is selected so as to observe the sediment wave dynamics over a long spatial scale.

5.2 Suspended sediment wave dynamics

Figure 2 shows the longitudinal profile of suspended sediment and the hydrodynamic waves at different time steps. At the initial stage of the release the two waves are in phase (Fig. 2a) and as the waves travel downstream they start to separate (Figs. 2b,c) with the separation increasing further downstream and finally the two waves are completely separated (Fig. 2d). Such physical phenomenon was also observed on field investigation of sediment flushing operation of the Cachí Reservoir in Costa Rica (Brandt 2005). Figure 2 also shows that the deposition (red line and is multiplied by 20 for visualisation purpose) is observed to be enhanced as the waves separate and most of the deposition is observed at the tail of the hydrodynamic wave. At the same time concentration wave decreases at the point of deposition. This shows that the interaction between the two waves significantly controls the deposition of the suspended sediments.

5.3 Effect of wave celerity factor

5.3.1 Wave celerity factor
The relationship between the celerity factor and selected hydraulic variables (depth, discharge, Rouse number, and relative roughness) are analysed for five different grain sizes. Figure 3 shows the plots of the relationships of depth, discharge, Rouse number, and relative roughness with α. The calculated range of variation of α applied in the model is between 0.93 to 0.97 (shown in Fig. 3 for $d_{50} = 230\ \mu m$). There is high dependence of α on flow depth and discharge for a given sediment grain size (Figs. 3a,b). As the flow depth or discharge increases the celerity factor generally increases regardless of the grain size. The celerity factor for fine grains is higher since the Rouse number is relatively small as a result giving a more uniform

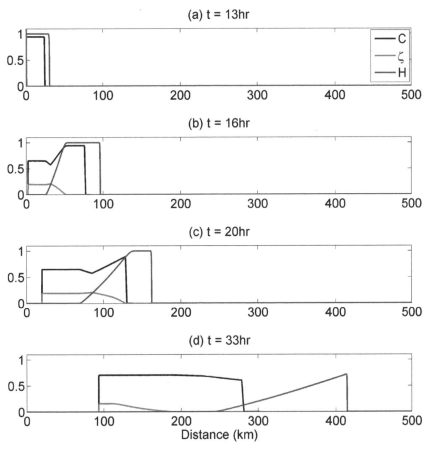

Figure 2. Sediment and hydrodynamic waves at various time steps: relative depth $(H–H_0)/(H_p–H_0)$, relative concentration $(C–C_0)/(C_p–C_0)$ and relative depth of deposition $20*(\zeta–\zeta_0)/(\zeta_p–\zeta_0)$ where ζ_p, H_p, and C_p = peak depth of deposition, peak flow depth, and peak concentration and ζ_0, H_0, and C_0 = minimum depth of deposition, base flow depth, and base concentration respectively.

concentration profile. For increasing Rouse number the celerity factor decreases regardless of the sediment grain size (Fig. 3c). The relationship between relative bed roughness and the celerity factor (Fig. 3d) is similar to the case of Rouse number such that as the relative bed roughness increases the celerity factor decreases regardless of the grain size. We also investigated the computed celerity factor for two different description of the reference height (z_r). We found that the ranges of variation of the computed celerity factor α with depth and discharge for the different grain sizes remained the same for $z_r = 2d_{50}$ and $z_r = 3d_{90}$. Thus, defining the reference height (z_r) as $2d_{50}$ (e.g. Zyserman & Fredsoe 1994) or as $3d_{90}$ (e.g. Van Rijn 1984) does not affect the estimated value of α.

5.3.2 Sediment concentration wave dynamics

Figure 4 shows the longitudinal profile of suspended sediment and the hydrodynamic waves at different time steps for two cases where the first is when the sediment wave celerity is corrected and the second is with the regular assumption that the sediment wave travels with average flow velocity $(\alpha = 1)$. In the latter case (shown in Fig. 4 in broken lines) there is less deposition comparatively since the wave separation is less enhanced. The region and length of deposition is also relatively different. There is less deposition when $\alpha = 1$ as a result the local concentration becomes higher than the one corrected by the celerity factor. This agrees well with two-dimensional (2-D) model study of Huybrechts et al. (2010). We evaluated the

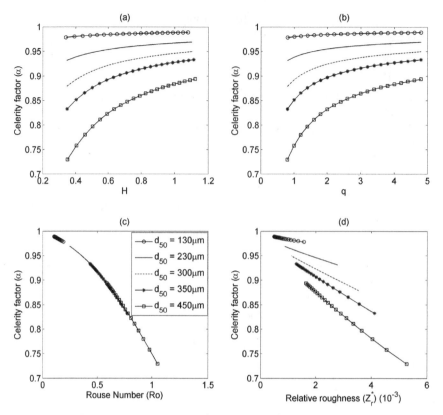

Figure 3. The relationship celerity factor with depth (H), discharge (q), Rouse number (Ro), and relative roughness (z_r^*).

mass balance at the end of simulation for the two cases: with and without the celerity factor. We observed that the proportion of boundary inputs of sediments that deposit were 10% and 7% and remain in suspension are 90% and 92% without and with celerity factor respectively. But we recognize that these figures may not lead us to a general conclusion as the computed percentages depend on the length of the domain as well as the duration of simulation. It is observed (Figs. 4c,d) that without celerity factor the suspended load is over predicted compared to the case with $\alpha < 1$. Huybrechts et al. (2010) implemented the celerity factor in 2-D model and compared simulated suspended load with the result of three-dimensional (3-D) model simulation and they obtained acceptable agreement with observed one when the celerity factor is considered. The effect of the celerity factor is also more evident and higher on the suspended load than on the concentration since the estimation of the suspended sediment load is directly related to the celerity factor. Over all, the sediment wave celerity affects the deposition of the fine sediments and its proper implementation is expected to improve the modelling of the deposition and transport of suspended sediments.

5.3.3 Sediment concentration and load behaviour

Figures 5a,b show the longitudinal pattern of modelled sediment concentration and loads respectively at $t = 20$ hr of the simulation with and without celerity factor when there is no bed exchange ($E - D = 0$). When the celerity factor is applied there is increment of sediment concentration, above the amount of concentration applied at the boundary, at the tail of concentration wave and at the tail of the hydrodynamic wave. When the celerity factor is not applied ($\alpha = 1$) such local increase in concentration is not evident at all. As shown in Equation 9, the last term of the equation contributes to this behaviour in concentration. But this term as well as the celerity factor could be very relevant in modelling suspended sediment

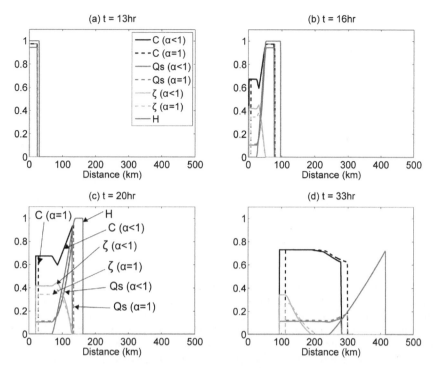

Figure 4. Sediment and hydrodynamic wave dynamics at various time steps with and without celerity factor: relative depth (H), relative concentration (C), and relative sediment load (Q_s) $(Q_s - Qs_0)/(Qs_p - Qs_0)$ where Qs_p and Qs_0 = peak and base sediment loads respectively.

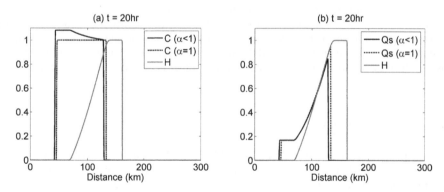

Figure 5. Sediment concentration (a) and sediment load (b) at $t = 20$ hr with and without celerity factor when there is no bed exchange.

transport. On the other hand, as shown in Figure 5b the sediment load shows a decrease at its front with its peak lower than the case of $\alpha = 1$. Figures 6a,b show the time evolution of sediment load and concentration at two downstream stations ($x = 90$ km, and 300 km from release point). With the celerity factor sediment load wave shows some decrease at its front and at the front of the hydrodynamic wave (Fig. 6a) though it agrees well with that of $\alpha = 1$ at the tail of the hydrodynamic wave. The effect of the celerity factor on the local increment of sediment concentration is evident only on the rising limb of the hydrodynamic wave. When the two waves are fully separated (Fig. 6b) the local increment vanishes. Overall, the local sediment concentration increase due to the celerity factor does not increase the sediment load.

171

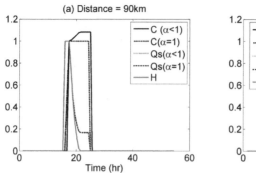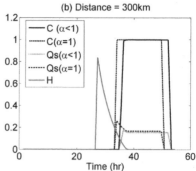

Figure 6. Sediment concentration and sediment load with and without celerity factor when there is no bed exchange at $x = 90$ km (a) and 300 km (b).

6 CONCLUSIONS

We studied how the interaction between the waves of suspended sediments and liquid discharge is controlled by the sediment wave celerity during reservoir sediment flushing operation. The effect of the sediment wave celerity on wave interaction and depositional processes has been investigated at larger spatial scale. Preliminary results of this study shows that the wave celerity controls the wave interaction significantly and thus the traditional assumption of representing sediment wave celerity as equal to the average flow velocity may lead to under-prediction of sediment deposition. We observed that the celerity factor enhances the wave separation and more deposition can be expected. We also investigated the dependence of the celerity factor on flow depth, discharge, Rouse number and relative roughness. There is generally an increasing trend in celerity factor with depth and discharge, while a decreasing trend is observed with Rouse number and relative roughness. The results presented in this study are based on the assumption that there is no sediment supply from the bed and only sediments supplied from the upstream are entrained and transported. The effect of changes of bed slope and channel geometry and dimensions were not taken in to account in this first analysis. The model also assumed the suspended sediments transported are uniform in grain size.

ACKNOWLEDGEMENTS

This work has been carried out within the SMART Joint Doctorate (Science for the Management of Rivers and their Tidal systems) funded with the support of the Erasmus Mundus programme of the European Union.

REFERENCES

Brandt, S. & Swenning, J. 1999. Sedimentological and geomorphological effects of reservoir flushing: The Cachí reservoir, Costa Rica. *Geografiska Annaler* 81: 391–407.

Brandt, S.A. 2005. Conceptualization of hydraulic and sedimentary processes in downstream reaches during flushing of reservoirs. In *Proceedings of the XXXI IAHR Congress, Water engineering for the future: choices and challenges, Seoul, Korea, 11–16 September 2005.*

Cao, Z. & Carling, P. 2002. Mathematical modelling of alluvial rivers: reality and myth. Part 1: General review. In *Proceedings of the ICE-Water and Maritime Engineering* 154: 207–219.

Celik, I. & Rodi, W. 1988. Modelling suspended sediment transport in nonequilibrium situations. *Journal of Hydraulic Engineering* 144(10): 1157–1191.

Duan, J. & Nanda, S. 2006. Two-dimensional depth-averaged model simulation of suspended sediment concentration distribution in a groyne field. *Journal of Hydrology* 327: 426–437.

Einstein, H.A. 1950. *The bed load function for sediment transportation in open channel flows.* U.S. Department of Agriculture, Soil Conservation Service: Washington, D.C.

Espa, P., Castelli, E., Crosa, G., Gentili, G. 2013. Environmental effects of storage preservation practices: Controlled flushing of fine sediment from a small hydropower reservoir. *Environmenmental Management* 52: 261–276.

Fischer, H., List, E., Koh, R., Imberger, J., Brooks, N. 1979. *Mixing in Inland and Coastal Waters.* Academic Press: New York.

Huybrechts, N., Villaret, C., Hervouet, J-M. 2010. Comparison between 2D and 3D modeling of sediment transport:application to the dune evolution. In Dittrich, Koll, Aberle and Geisenhaainer(eds.), *Proceedings of the 5th IAHR International Conference on Fluvial Hydraulics, Braunschweig, Germany, 8–10 September 2010.*

Liu, J., Minami, S., Otsuki, H., Liu, B., Ashida, K. 2004. Environmental impacts of coordinated sediment flushing. *Journal of Hydraulic Research* 42: 461–472.

Rathburn, S. & Wohl, E. 2001. One-dimensional sediment transport modeling of pool recovery along a mountain channel after a reservoir sediment release. *Regulated Rivers: Research and Applications* 7(3): 251–273.

Richardson, J.F. & Zaki, W.N. 1954. Sedimentation and fluidization. Part I. *Transactions of the Institution of Chemical Engineers* 32(1): 35–53.

Siviglia, A. & Toro, E.F. 2009. WAF method and splitting procedure for simulating hydro- and thermal-peaking waves in open-channel flows. *Journal of Hydraulic Engineering* 135(8): 651–662.

Toffolon, M., Siviglia, A., Zolezzi, G. 2010. Thermal wave dynamics in rivers affected by hydropeaking. *Water Resources Research* 46.

Toro, E.F. 1989. A weight average flux method for hyperbolic conservation laws. In *Proceedings of the Royal Society of London, Mathematical and Physical Sciences* 423: 401–418.

Van Rijn, L. 1984. Sediment transport. Part II: Suspended load transport. *Journal of Hydraulic Engineering* 110: 1613–1641.

Wohl, E. & Rathburn, S. 2003. Mitigation of sedimentation hazards downstream from reservoirs. *International Journal of Sediment Research* 18: 97–106.

Wright, S. & Parker, G. 2004. Flow resistance and suspended load in sand-bed rivers: Simplified stratification model. *Journal of Hydraulic Engineering* 130: 796–805.

Zolezzi, G., Siviglia, A., Toffolon, M., Maiolini, B. 2010. Thermo-peaking in alpine streams: event characterization and time scales. *Ecohydrology* 4: 564–576.

Zyserman, J.A. & Fredsoe, J. 1994. Data analysis of bed concentration of suspended sediment. *Journal of Hydraulic Engineering* 120: 1021–1042.

Reservoir Sedimentation – Schleiss et al. (Eds)
© *2014 Taylor & Francis Group, London, ISBN 978-1-138-02675-9*

Numerical modeling of suspended sediment transport during dam flushing: From reservoir dynamic to downstream propagation

G. Antoine
EDF R&D LNHE & LHSV, Chatou, France

A.-L. Besnier & M. Jodeau
EDF R&D LNHE, Chatou, France

B. Camenen
Irstea, Lyon, France

M. Esteves
IRD LTHE, Grenoble, France

ABSTRACT: Dam flushing is often performed to remove sediments of all sizes from reservoirs while minimizing downstream impacts. In this study, the reservoir flushing operation and the downstream transport of suspended sediments are simulated based on a numerical tool. The study site is the Arc River in the French Alps, focusing on two reaches: locally on the Saint Martin la Porte (SMLP) Reservoir and a few metres upstream (reach 1), and downstream the 120 km between the SMLP reservoir and the city of Grenoble (reach 2). The chosen numerical code is COURLIS, developed at EDFRD, and which could be coupled with codes of the open source Telemac-Mascaret system. The model was applied to the 2012 flushing event of the Arc River dams, which are managed by EDF (Electricity of France). The bed erosion and the suspended sediment transport dynamics in the most downstream reservoir (SMLP) were first computed. The initial state of the deposited fine sediment were defined using measured topographic data of the reservoir. The upstream boundary condition for the reach 2 model was set from estimated discharges and suspended sediment concentrations. The downstream study reach is approximately 120 km long, from SMLP dam to the city of Grenoble along the Arc and Isere rivers. Topographic data of the river bed were used to define the downstream river reach, taking into account the vegetated alternate bars. Water discharges and suspended sediment concentrations were also measured at several locations in the rivers downstream of the flushed reservoirs. A new formula is proposed to estimate two mean concentration values in the cross section: one for the main channel and the other for the overbank section. A good agreement with field data is obtained for both the reservoir and downstream reach dynamics. Moreover, the key role of the alternate bars in the fine sediment dynamics is highlighted, quantified and discussed.

1 INTRODUCTION

Dam flushing is an efficient way to manage sediment accumulation in reservoirs [17, 21]. In the case of gate structure dams, flushing operations could be performed in order to have a quasi-natural flow in the reservoir characterized by high velocities and low water depths. These conditions could allow the removal of significant quantities of fine sediments and gravels.

However, the downstream propagation of the released suspended sediments could strongly impact the river bed morphology, as well as the river system ecology [4, 8, 9, 18]. Therefore, flushing operations should be managed in order to mitigate the downstream impacts.

Numerous studies have been undertaken in order to analyze and simulate the hydro-sedimentary processes in reservoirs and in the downstream river. Few of these studies concern dam flushing events, and even less directly quantify the downstream impact of the suspended sediment propagation.

Numerical modeling tools are quantitative management means to predict erosion, transport and deposition of sediment due to dam flushing. To assess their validity they should be compared with measured in situ data. A few studies present such results [15, 2, 5, 22, 20]. Furthermore there are very few works that have simultaneously investigated the global system of the reservoir and the downstream river reach. Numerical tools could be a convenient tool to test sediment management strategies.

We propose herein a numerical modeling of a real dam flushing event, performed to study both the reservoir dynamics and the downstream propagation of suspended sediment. The first part of this paper describes the study site and the sediment management operations, next the numerical code is presented. Results of the numerical modeling are then presented for the reservoir area and in the downstream river reach.

2 FINE SEDIMENT MANAGEMENT IN THE ARC AND ISERE RIVERS

The Arc River is located in the northern French Alps, its watershed has an area of 1950 km² and the geology and mountains slopes induce high rates of both fine sediment and gravels transport in the river. As an order of magnitude, at the Arc river outlet, the annual flux of fine sediment is estimated to be about one million tons and the mean bedload flux to be about 500 000 tons. The river bed is made of boulders and gravels with patches of fine sediments and vegetation. The river is diked on a large part, the mean width between dikes is about 50 m. The maximal slope of the river is 3%. The Arc River is the main tributary of the Isere River (watershed area at Grenoble city 5570 km²), which has a minimal slope of 0.1% and its width between dikes is 200 m.

Several hydroelectricity facilities had been built in these valleys. Figure 1 (a) shows the watershed of the Arc river and the main facilities. The reservoirs undergo a high rate of sedimentation. Therefore the three reservoirs located in the river Arc are flushed once a year. The three reservoirs are gated structure dams that can be totally opened. There are also 3 basins located on water diversions which allow flexibility in the electricity production and siltation control. The flushing of these basins is not effective, consequently another strategy including dredging is chosen to manage their sedimentation. This study focuses on the flushing of the 3 river reservoirs and in particular the sediment processes in the most downstream reservoir: the Saint Martin la Porte (SMLP) Reservoir.

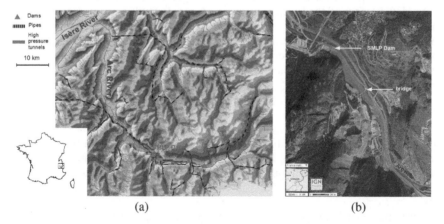

(a)   (b)

Figure 1.   (a) Location of the Arc watershed, and description of hydroelectricity facilities in the Arc river valley. (b) Aerial picture of Saint Martin la Porte (SMLP) reservoir (source: Geoportail, IGN).

The study area has already been the subject of other research works, which include works on gravel bar morphodynamics [12], measurement of fine sediment transport and settling velocities [1], hydrometry research [13, 6]... Therefore a comprehensive set of in situ data is available.

The work presented in this paper concentrates on the fine sediment management, though gravel management is also a major issue in the Arc river.

The 10 m high SMLP dam (Fig. 1 (b)) dam was constructed in 1986, and has a storage capacity of 68 000 m^3. The Arc drainage area is 1167 km^2 at the dam. The reservoir is located downstream of the Saussaz hydro power plant and has awater intake for the next hydro power plant (Hermillon). Both design flow rates are 90 m^3/s and the guaranteed flow discharge downstream of the SMLP dam is 7 m^3. Since the construction of the dam, several bathymetry and sediment sampling surveys allow to describe the evolution of the reservoir (see Fig. 1 (b) to identify the areas):

- in the downstream area, downstream the rock spur, there is a siltation area where fine sediment deposition occurs and where the annual flushing operations help to maintain the storage capacity;
- from the rock spur to the small bridge, there is an aggradation zone, where the flushing operations are not completely efficient, therefore there is a loss of storage capacity in this area;
- upstream of the bridge, there is a gravel deposition zone;

Sediments are sorted in the reservoir from gravels ($d_{50} = 6$ cm) in the upstream part to silt and sand just upstream the dam.

3 PRESENTATION OF THE NUMERICAL TOOL

One dimensional modelling is one of the different options for modeling flow and sediment transport in reservoirs, and is well suited for hydrosedimentary analyses if flow patterns in deep areas is not of major concern. One dimensional modeling is well suited if long term simulations are expected and if average values are adequate. The COURLIS numerical code allows the calculation of one dimensional flow and the sediment transport of mud and sand. The code has been used to compute flushing flow or emptying of several reservoirs [2, 16, 14], therefore it has demonstrated its relevance.

3.1 *Cross section view*

Despite a one dimensional computation for the flow, the sediment bed is described according a bidimensionnal view, that is to say sediment layers are defined transversally to the flow axis, Figure 2. Each sediment layer is characterized by its thickness, percentages of sand and silt, sand grain size and cohesive sediment characteristics.

3.2 *Equations*

COURLIS is based on a coupling between the component MASCARET which solves the 1D shallow water equations [10] and the sediment component which handles sediment processes.

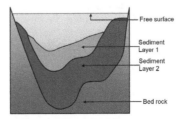

Figure 2.   Sediment layers in the cross section.

Sand and cohesive sediment are dealt withseparately. For both type of sediment an advection-dispersion is solved:

$$\frac{\partial AC}{\partial t} + \frac{\partial QC}{\partial x} = \frac{\partial}{\partial x}\left(kA\frac{\partial C}{\partial x}\right) + E - D \tag{1}$$

where A is the cross sectional flow area, C the suspended concentration, Q the flow rate, k the dispersion coefficient, E and D respectively the Erosion and Deposition linear rates.

3.2.1 Cohesive sediments

For **cohesive sediments**, local deposition or erosion processes are considered. In particular, the local bed shear stress is calculated laterally in each cross section as follow:

$$\tau(y) = \rho g h(y) J(y) \tag{2}$$

where ρ is the mass density of the water, g is the gravity, $h(y)$ the local water depth (variable in the section) and $J(y)$ the local friction loss. The local friction loss is calculated using the Manning-Strickler formula, with the mean hydraulic parameters estimated for the main channel (mc) or the overbank section (b), according to the Debord formula [19]:

$$\begin{cases} \text{if } y \text{ is in the main channel} & J(y) = J_{mc} = \dfrac{U_{mc}^2}{K_{mc}^2 R_{mc}^{4/3}} \\[3mm] \text{if } y \text{ is in the overbank section} & J(y) = J_b = \dfrac{U_b^2}{K_b^2 R_b^{4/3}} \\[3mm] \text{if } y \text{ is between both sub-sections} & J(y) = (J_{mc} + J_b)/2 \end{cases} \tag{3}$$

where U is the mean velocity, K Strickler coefficient and R the hydraulic radius.

To be consistent with the hydraulic variables, calculated for each sub-section of a compound channel system, an averaged value of the suspended concentration is estimated for the main channel (C_{mc}) and the overbank section (C_b). To calculate these two variables, two equations are used. The first equation is based on conservation of mass:

$$QC = Q_{mc} C_{mc} + Q_b C_b \tag{4}$$

The second equation is physically based on the estimation of the ratio η_C defined as:

$$\eta_C = \frac{C_b}{C_{mc}} \tag{5}$$

Based upon laboratory analyses, the data set obtained in two experiments in compound channels, Hu et al. (2010 [11]) expressed η_C as functions of mean velocities and water depths in each sub-section as follows:

$$\frac{C_b}{C_{mc}} = \alpha_U \left(\frac{U_b}{U_{mc}}\right)^{\beta_U} \tag{6}$$

and:

$$\frac{C_b}{C_{mc}} = \alpha_H \left(\frac{H_b}{H_{mc}}\right)^{\beta_H} \tag{7}$$

where (α_U, β_U) and (α_H, β_H) have been measured for two configurations. More precisely, two couples of rugosity values for the main channel and the overbank section have been tested in laboratory by Hu et al. [11]. For these two configurations (where $R \approx H$, the constant water depth in each sub-section), the ratio between the mean values of bed shear stress can be expressed as:

$$\frac{\tau_b}{\tau_{mc}} = \left(\frac{U_b}{U_{mc}}\right)^2 \left(\frac{K_{mc}}{K_b}\right)^2 \left(\frac{H_{mc}}{H_b}\right)^{1/3} \tag{8}$$

Using Equation 8, Equation 6 and Equation 7, we finaly expressed η_C as follow:

$$\eta_C = \frac{C_b}{C_{mc}} = \alpha_C \left(\frac{\tau_b}{\tau_{mc}}\right)^{\beta_C} \tag{9}$$

where:

$$\alpha_C = \left[\alpha_U^{2/\beta_U} \alpha_H^{-1/3\,\beta_H} \left(\frac{K_b}{K_{mc}}\right)^2 \right]^{\frac{3\beta_U\beta_H}{6\beta_H - \beta_U}} \tag{10}$$

and:

$$\beta_C = \frac{3\beta_U\beta_H}{6\beta_H - \beta_U} \tag{11}$$

The expression, given by Equation 9 allows to solve the system of equation Equation 4 and Equation 5.

For erosion and deposition processes, the Partheniades and Krone empirical formulae are used:

$$E(y) = M\left(\frac{\tau_{eff1}(y)}{\tau_{CE}} - 1\right) \tag{12}$$

$$D(y) = w_s C(y)\left(1 - \frac{\tau_{eff1}(y)}{\tau_{CD}}\right) \tag{13}$$

where M is a erosion rate coefficient, w_s a settling velocity, τ_{CD} and τ_{CE} deposition and erosion critical shear stresses. $\tau_{eff1}(y)$ is defined as $\tau(y)$ (Equation 2), but using a different friction coefficient K_{MUD}, which is a skin rugosity coefficient. The concentration value $C(y)$ depends on the position of y in the cross section: in the main channel $(C(y) = C_{mc})$, in the overbank section $(C(y) = C_b)$ or between the two sections $(C(y) = (C_{mc} + C_b)/2)$.

3.2.2 Sand

For **sand**, *i.e.* non cohesive sediments, the transport capacity is calculated with the Engelund Hansen [7] formula, using average values of hydraulic parameter in the cross section:

$$q_s = 0.05\sqrt{\frac{\delta d^3}{g}} \frac{K^2 R^{1/3}}{(\rho_s - \rho)gd} \tau_S \quad \text{where } \tau_{eff2} = \tau\left(\frac{K}{K_{SAND}}\right)^{3/2} \tag{14}$$

where d is the represenative diameter of the suspended sediments, and K_{SAND} is a grain friction coefficient and could be written $K_{SAND} = 26/d_{90}^{1/6}$.

It allows us to define an equilibrium concentration, C_{eq}:

$$C_{eq} = \frac{\rho_s q_s}{Q} \tag{15}$$

179

and the erosion an deposition rates:

$$\begin{cases} \text{if } C_{sand} \geq C_{eq} \text{ deposition} & D = w_s \left(C_{sand} - C_{eq} \right) \\ \text{if } C_{sand} \leq C_{eq} \text{ erosion} & E = w_s \left(C_{eq} - C_{sand} \right) \end{cases} \quad (16)$$

3.2.3 *Bed evolution*
The bed evolution depends on erosion and deposition rates

$$\frac{\partial Zb}{\partial t} = \frac{D}{C_{deposition}} - \frac{E}{C_{layer}} \quad (17)$$

The bank failure is taken into account using a simple model, the bank slope is compared to a stability slope (submerged or emerged). If the critical slope is exceeded, sediment deposit is supposed to collapse immediately and is put in suspension in the flow.

3.3 *Splitting*

The numerical code uses a splitting approach to solve hydraulic and sediment transport equations. Both hydraulic and sediment components could be coupled at each time step, *i.e.* the hydraulic variables are calculated for a fixed bed then the bed evolution is calculated. If the hydrodynamic varies slowly, the user could define a less frequent coupling, for example coupling every ten or more hydraulic time steps.

4 MODELING OF 2012 FLUSHING EVENT

4.1 *Flushing of Saint Martin la Porte Reservoir*

A one dimensional model is appropriate to reproduce sediment dynamic during a flushing event because the flow is mainly directed in the downstream direction. Observations during the event do not highlight major transverse currents. A 2d or 3d model should be more appropriate in the study of the siltation in the reservoir.

A specific model was built to reproduce sediment processes in the SMLP reservoir. According to field surveys, the bed composition has been simplified to enable a description by several sediment layers. The structure of the 1D model is described by three sedimentary components, which are depicted in Figure 3. The upstream bed made of gravel is reproduced with a rigid bed because gravel processes are not the focus of this study. The first component is a 90 cm sand layer followed downstream by a second sedimentary component of a 1 m mixed sand and silt layer. The first two components are covered by a 50 cm silt layer. The non erodible rigid layer was defined according to a 2008 bathymetry, because a flood in 2008 is assumed to have eroded the major part of sediment in the reservoir.

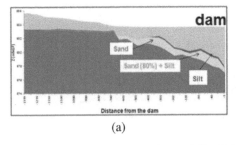

(a)   (b)

Figure 3.   (a) Sedimentary layers for modeling Saint-Martin-la-Porte Reservoir. (b) Downstream view of SMLP during a flushing event.

The geometry of the model is based on a multi beam bathymetric survey performed a few weeks before the 2012 flushing, a 1200 m long stretch of the reservoir is modeled. Modeling results are compared with a bathymetric survey performed a few weeks after the flushing.

The mesh size is 5 m in the domain except in the 60 m upstream the dam where the mesh is finer (2 m). The geometry has been extended by a 200 m reach at the upstream side to avoid boundary effect and a 3 m reach is added at the downstream side to well reproduce the dam effect.

The upstream boundary condition is a measured discharge and a measured sediment concentration. The water level at the downstream boundary condition is defined according to the operating conditions, the downstream condition for the sediment concentration is a Neumann boundary condition with $\partial C/\partial x = 0$. Some tests have been performed to take into account the different opening of the dam gates, the downstream section has been modified in function of the number of opened gates.

The friction coefficient is 40 m$^{1/3}$s^{-1} except on the dam concrete sill where a value of 60 m$^{1/3}$s^{-1} is chosen.

The numerical results are compared with measurements at three particular points:

- water level using laser scan measurements performed during the event;
- bed evolutions using bathymetries;
- concentrations at the dam.

There is a good agreement between both on the three previous variables. Figure 4 shows the comparison between measured and simulated sediment concentrations.

4.2 Suspended sediment transport in the downstream Arc and Isere Rivers

For the downstream part of the river reach, the mesh size is larger than in the reservoir part (about 300 m). The hydraulic part of the model has been calibrated with two other flushing events, in 2010 and 2011. Six homogeneous reaches have been defined to calibrate the friction coefficients. During the 2011 flushing event, the measured value of the base-flow discharge was low on the Isere River, as well as the overbank sections of the river bed haven't be flooded. The Strickler coefficient corresponding to the main channel has been well calibrated with this event. During the 2010 flushing event, the overbank sections have been widely flooded, because of the high discharge values in the Isere River. With this event, the Strickler coefficients of the banks have been successfully calibrated. The resulting values of Strickler coefficient used to simulate the 2012 flushing event are presented in the Table 1.

The bed composition of the river reach is defined by one sediment layer. The thickness of this unique sediment layer is defined as a function of the local slope and the lateral position

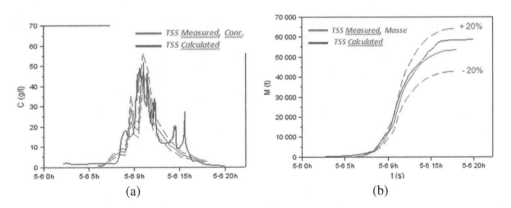

(a)　　　　　　　　　　　　(b)

Figure 4. Comparison between numerical results and measurements at SMLP dam: (a) Sediment concentration. (b) Cumulative mass.

Table 1. Strickler coefficients used along the downstream river reach for the main channel and the overbank sections.

	Min dist from the dam (km)	Max dist from the dam (km)	K_{mc} (m$^{1/3}$/s)	K_b (m$^{1/3}$/s)
Reach 1	0	9	25	10
Reach 2	9	15	27	27
Reach 3	15	21	30	30
Reach 4	21	41	30	15
Reach 5	41	51	40	10
Reach 6	51	114	45	10

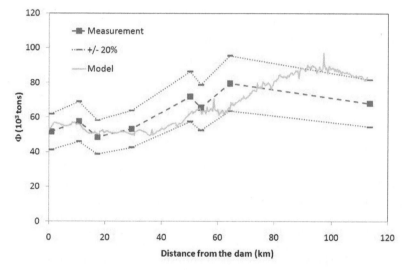

Figure 5. Total suspended mass transported during the flushing event dowstream the dam. Comparison between numerical results (green) and measurements (blue).

in the cross section: more erodible mass of fine sediment is available on the overbank sections, and more erodible mass is available in reaches with small slope values.

Concerning the values of sediment parameters, most of them have been measured. The erosion parameters for the Parthniades law have been defined as $\tau_{CE} = 5$Pa and $M = 10^{-3}$ kg/m^2/s. The parameters of the Krone's law have been fixes as $\tau_{CD} = 0.8$ Pa, and the settling velocity have been measured in situ during the 2012 flushing event and is equal to $w_s = 0.8$ mm/s. The skin friction coefficient is equal to $K_{MUD} = 85$ m$^{1/3}$/s [3], and the value of the dispersion coefficient have been measured on the Arc River in 2013, and is equal to $k = 70$ m^2/s.

The estimated discharges and suspended sediment concentrations from the reservoir part of the model are directly used as an upstream limit condition of the downstream river reach model. The downstream condition for the sediment concentration is a Neumann boundary condition with $\partial C/\partial x = 0$. Finally, one sediment class has been considered for the downstream propagation. The results of the numerical model are presented in the Figure 5. In order to compare numerical results with measured data, the total suspended sediment fluxes $\Phi = \int_{T_{event}} Q(t) \times C(t)dt$ have been estimated at each measurement site along the river reach.

The Figure 5 shows that the model reproduces the measured data with a good agreement. The total erosion observed along the river reach is well reproduced, as well as the deposition dynamic on the vegetated banks. Particularly, the numerical results don't overestimate deposition on the overbank sections thanks to the new implemented formulation.

5 DISCUSSION

We show in this study that the suspended sediment dynamic in the reservoir could be well reproduced with the COURLIS model, as well as the downstream propagation along a 120 km river reach. In both cases, the river bed composition has been defined precisely, with strong physical meaning. Measurements of suspended concentration have been used to validate the numerical results. However, a very poor description is available concerning the interaction between mud and sand fractions during transport. A good definition of the sand layers was necessary to well reproduce the local erosion dynamic into the reservoir. But no precise information concerning the downstream propagation of sand is available. This kind of measurement would help to get better results, for example by taking into account mud and sand interactions.

6 CONCLUSION AND PERSPECTIVES

In this study, the dam flushing performed in 2012 on the Arc River has been simulated. The suspended sediment dynamics in the reservoir has been well reproduced by COURLIS, thanks to a detailed description of the sediment layers. A new formulation is presented to take into account the suspended sediment transport in the compound channel system, and applied to the downstream river reach. The model successfully reproduced field measurements, using as upstream limit condition the results from the reservoir part. As perspectives, a better investigation of the sand transport dynamic and propagation would be helpful to improve numerical results. A complete coupling between the reservoir and downstream models will also be useful, in particular for the testing of different flushing strategies.

REFERENCES

[1] G. Antoine, M. Jodeau, B. Camenen, and M. Esteves. A settling velocity parameterisation for sand/mud mixture in a 1d flow during a flushing event. In *Proceedings of the River Flow Conference*, 2012.

[2] C. Bertier, J.-P. Bouchard, and L. Dumond. One dimensional model for reservoir sedimentation management. In *River Flow*, 2002.

[3] J.-P. Bouchard and C. Bertier. Morphological change in reservoirs in relation to hydraulic conditions. In *River Flow*, 2008.

[4] S.A. Brandt. Sedimentological and geomorphological effects of reservoir flushing: The Cachi reservoir, Costa Rica, 1996. *Geografiska Annaler A*, 81:391–407, 1999.

[5] H. Chang, L. Harrison, W. Lee, and S. Tu. Numerical modeling for sediment-pass-through reservoirs. *Journal of Hydraulic Engineering*, 122-7, 1996.

[6] G. Dramais, J. Le Coz, B. Camenen, and A. Hauet. Advantages of a mobile lspiv method for measuring flood discharges and improving stage-discharge curves. *Journal of Hydro-Environment Research*, 5-4:301–312, 2011.

[7] F. Engelund and E. Hansen. A monograph on sediment transport in alluvial streams. *Technical University of Denmark*, 1967.

[8] G. Fasolato, P. Ronco, and Y. Jia. Studies on sediment transport and morphodynamic evolution of a river due to sediment flushing operations of an alpine reservoir. In *iahr 2007*, 2007.

[9] F. Gallerano and G. Cannata. Compatibility of reservoir flushing and river protection. *Journal of hydraulic Engineering*, 137-10:1111–1125, 2011.

[10] N. Goutal and F. Maurel. A finite volume solver for 1D shallow-water equations applied to an actual river. *Int. J. Numer. Meth. Fluids*, 38:1–19, 2002.

[11] C. Hu, Z. Ji, and Q. Guo. Flow movement and sediment transport in compound channels. *Journal of Hydraulic Research*, 1:23–32, 2010.

[12] M. Jaballah, B. Camenen, A. Paquier, and M. Jodeau. 2-d numerical modeling of water flow over a gravel bar. In *River Flow*, 2012.

[13] M. Jodeau, A. Hauet, A. Paquier, J. Le Coz, and G. Dramais. Application and evaluation of LS-PIV technique for the monitoring of river surface velocities in high flow conditions. *Flow Measurement and Instrumentation*, 19:117–127, 2008.

[14] M. Jodeau and S. Menu. Sediment transport modeling of a reservoir drawdown, example of Tolla reservoir. In *River Flow Conference*, 2012.

[15] A. Khosronejad, C.D. Rennie, A.A. Salehi Neyshabouri, and I. Gholami. Three-dimensional numerical modeling of reservoir sediment release. *Journal of Hydraulic Research*, 46:209–223, 2008.

[16] D. Marot, J.-P. Bouchard, and A. Alexis. Reservoir bank deformation modeling: application to grangent reservoir. *Journal of Hydraulic Engineering*, pages 586–595, 2005.

[17] G. L. Morris and J. Fan. *Reservoir Sedimentation Handbook*. McGraw-Hill, 1997.

[18] U. Mrle, J. Ortlepp, and M. Zahner. Effects of experimental flooding on riverine morphology, structure and riparian vegetation: the River Spl, Swiss National Park. *Aquatic Science*, 65:191–198, 2003.

[19] G. Nicollet and M. Uan. Ecoulements permanents surface libre en lits composs. *La Houille Blanche*, (1):21–30, January 1979.

[20] S.L. Rathburn and Wohl E.E. One-dimensional sediment transport modeling of pool recovery along a mountain channel after a reservoir sediment release. *Regulated Rivers: Research & Management*, 17:251–273, 2001.

[21] H.W. Shen. Flushing sediment through reservoirs. *Journal of Hydraulic Research*, 37–6:743–757, 1999.

[22] E. Valette, P. Tassi, M. Jodeau, and C. Villaret. St Egrve Reservoir—Multi-dimensional modelling of flushing and evolution of the channel bed. In *River Flow*, 2014.

Reservoir Sedimentation – Schleiss et al. (Eds)
© 2014 Taylor & Francis Group, London, ISBN 978-1-138-02675-9

St-Egrève reservoir—modelling of flushing and evolution of the channel bed

E. Valette
Centre d'Ingénierie Hydraulique—Electricité De France, Le Bourget du lac, France

P. Tassi & M. Jodeau
Laboratoire National d'Hydraulique et Environnement—Electricité De France, Chatou, France

C. Villaret
HR Wallingford, Wallingford, UK

ABSTRACT: The Saint-Egrève dam is located downstream of the city of Grenoble, on the Isere river in France. Over time, the reservoir has silted up. Frequent flushing operations allow the maintenance of a channel in the reservoir, but siltation bank formation on either side of the channel is irreversible. Due to the urban location of the reservoir, maintaining the freeboard of the upstream dike of the reservoir during the design flood is a major issue. Nowadays, the evolution of the filling is such that the channel erosion during the flood must be taken into account to estimate a realistic freeboard. During a first part of the study, 1D morphodynamics simulations were performed. The model was calibrated and validated with measured sediment fluxes and then applied to a project design situation. A second part of the study consisted on performing 2D and 3D numerical simulations and comparing results with the 1D model.

1 INTRODUCTION

Sediment transport and deposition in reservoirs are natural processes. Recently, it has been estimated that the worldwide sedimentation in reservoirs corresponds to about one per cent of the whole capacity per year (Mahmood 1987). In specific areas, sedimentation rates can be significantly higher; it may reach more than 70% of reservoir initial capacity (Bouchard & Bertier 2008). The filling of reservoirs depends on the production of sediment from the watersheds, the hydrology of the watershed, geometry and hydraulics of the reservoir and management of reservoir capacity (Morris 1997). The reservoir sedimentation impacts the river reach upstream the dam as much as the downstream reach: storage loss, delta deposition, blocking or clogging of intakes or bottom gates, downstream erosion, ecology, etc. Consequently, one has to take into account sediments when operating dams; therefore we need means to predict the consequences of dam operations on sediment transport and reservoir morphology.

Flushing operations aim at eroding sediments from reservoirs to maintain or to increase their storage capacity and/or prevent flooding upstream the dam. In such operations, the release of sediments to the downstream reach may be significant and should be controlled (Brandt 2000). There are different ways of predicting the downstream impacts of such operations, often relying on the experience. Nevertheless, numerical modeling can be used as a tool for planning and operation activities. The choice of the numerical tool depends on the available data, the geometry of the reservoir, the operating conditions and the objectives of the simulation. In this work, the flushing of Saint Egrève reservoir is simulated with different modules of the TELEMAC-MASCARET System (www.opentelemac.org) and comparisons with a comprehensive set of experimental data are presented and discussed.

2 SITE DESCRIPTION

The St-Egrève reservoir (France) is located in the Grenoble urban area, as shown on Figure 1, downstream the confluence of the Isère and Drac rivers (9270 km^2 catchment area). The St-Egrève dam is a run-of-river power station, with a maximum turbine discharge of 540 m^3/s. The dam comprises 5 identical openings with overflow flaps, and a 25-meter wide tainted gate with 6 meters of lifting height and a weir at elevation 196.50 m NGF. The normal reservoir level (FSL) during operation is 205.50 m NGF. The capacity of the reservoir in 1992 was 3.86 hm^3 from the dam to the confluence. For safety reasons, a security distance of 1 meter with respect to the crest of the reservoir embankment must be guaranteed for a flood of 3000 m^3/s.

The Isère River is highly loaded with fine sediment, and in the St-Egrève reservoir this sediment is deposited. The St-Egrève reservoir shows a sediment accumulation on the left bank that continues to silt up (see Fig. 2). In 2010 its elevation was 204.5 m NGF on average, i.e. one meter below the FSL. If this bar continues to silt up, bank volume will grow from 1 hm^3 presently to 1.45 hm^3. The remaining channel has a variable topography in its cross sections: its minimum area in the absence of flushing can be estimated at 250 m^2, i.e. a volume of about 0.6 hm^3 along 2500 meters. The channel is deepened during floods with flushing operations, and the maximum reservoir capacity that can be reached is estimated at 3 hm^3, as shown on Figure 2.

Figure 1.   Location of the St-Egrève reservoir.

Figure 2.   St-Egrève reservoir during a flushing event and evolution of the reservoir capacity.

186

3 1D MORPHODYNAMICS SIMULATIONS

3.1 *Aim of the study*

The COURLIS software (internally coupled to the 1D hydrodynamics model MASCARET) was used to determine the bottom evolution kinetics during floods. COURLIS computes the bottom evolutions in a channel section as a function of the bed shear distribution in the cross-wise direction of the flow (Marot et al. 2005, Jodeau & Menu 2012). This numerical code deals with cohesive and non cohesive sediment. The objective was to verify that when starting with a high degree of siltation, the erosion at the start of flushing is sufficient to guarantee the preservation of the 1-meter freeboard with respect to the crests of the dikes. In this work, only the calibration and validation of the model are presented.

3.2 *Available measurements*

Two events were used to calibrate and validate the numerical model: flushing operations performed in May 2008 and May-June 2010. Bathymetries of the reservoir were surveyed before and after each of these events. Turbidity meters placed upstream and downstream of the reservoir enable monitoring of the sediment concentration evolution during a flood and determination of the silt volumes which passed through between the bathymetric measurements and the flushing operations. In addition, sediment samples were taken from the reservoir in September 2010.

The calibration data corresponds to the May 2008 flushing operation. This flushing operation was preceded and followed by two bathymetric measurements, one in April and one in August 2008. The grid was based on cross-section profiles 100 meters apart derived from the bathymetric data. These profiles were pre-processed to achieve a calculation profile every 50 meters. The chosen Strickler coefficient is taken equal to 45 $m^{1/3}$ s^{-1}.

Model calibration revealed the necessity to model three distinct sediment layers, as shown in Figure 3. The top layer represents the slightly consolidated sediment (easily remobilized), the second layer the recently deposited sediment (few years), and the third sediment layer the most consolidated. Sediment layers were constructed from the bathymetric data: (i) In

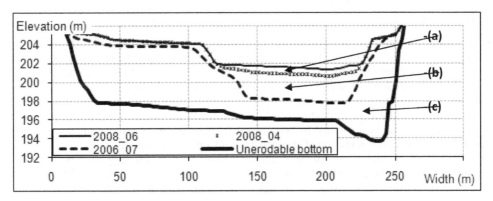

Sediment layers	Concentration (kg/m³)	Critical erosion shear stress (Pa)	Erosion parameter (kg/s/m²)
(a) Slightly consolidated	1000	1	0.005
(b) Recent	1100	6	0.06
(c) Old	1100	8	0.02

Figure 3.   Sedimentary layer creation: transversal view and sediment characteristic.

2008, the previous flushing operation dated from April 2006. The layer of old sediment was comprised between the non-erodible bottom measured in the topographic surveys prior to the impounding of the dam and the bathymetry of July 2006; (ii) The layer of recent sediment was represented by the sediment deposited between the bathymetries of July 2006 and April 2008; and (iii) The layer of slightly consolidated sediment was considered equal to the estimated volume of sediment deposited between the date of the bathymetry and the flushing episode, i.e. 150,000 tons. Assuming a dry matter concentration of 1000 g/l, this layer represents a deposit of roughly 70 cm at the bottom of the channel with respect to the April 2008 profile. Besides, the slope stability angle is considered constant and equal to 15°.

The erosion shear stress parameter τ_{eff} used in COURLIS is defined as:

$$\tau_{eff} = \tau \cdot \left(\frac{K}{K_p}\right)^2 = \rho g h \cdot \frac{U^2}{K^2 R_h^{4/3}} \left(\frac{K}{K_p}\right)^2 = \rho g h \cdot \frac{U^2}{K_p^2 R_h^{4/3}} \tag{1}$$

where τ = total shear stress (Pa), K = Stickler coefficient (m$^{1/3}$ s^{-1}), Kp = Skin Strickler coefficient (m$^{1/3}$ s^{-1}), ρ = water density (kg/m^3), g = gravity (m/s^2), h = local water depth (m), R_h—hydraulic radius (m), U = mean velocity in the cross section (m/s).

Using hydraulic radius instead of local water depth is an alternative given in the software, which was adopted in this study (after calibrations tests). Therefore, the equation (1) becomes:

$$\tau_{eff} = \rho g \cdot \frac{U^2}{K_p^2 R_h^{1/3}} \tag{2}$$

The flushing parameters and the results are summarized on Figure 4. The following phases can be observed:

- During the lowering of the water level (phase 1), we observe an erosion peak due to the slightly consolidated silt ①,
- The main erosion peak ② corresponds to the end of the phase 1, when the water level has reached its minimum level. It is well represented by the 1D model although the maximum peak value is slightly underestimated, The increased erosion ③ is due to the passage of the flood peak (phase 3). This third peak is also underestimated. ③

The calculated mass of eroded sediment (1.13 Mt) is in good agreement with the measured mass (1.14 Mt). The evolution of the eroded mass over time can also be well adapted. At the time of the flood peak on May 30 at 1 pm, over a million tons of sediment had already been eroded. A comparison of the profiles measured during August 2008 with the computed profiles yields to satisfactory results, Figure 5.

However, the COURLIS simulation results in the channel are deeper than those measured in bathymetry. This can be attributed to a considerable accumulation of sediment in the reservoir during the period between the end of flushing (end of COURLIS simulation) and the bathymetry date. Indeed, the high flow episode of June 2008 lasted after the flushing, causing solid matter inflows and significant settling in the reservoir of an order of magnitude of 300,000 tons (i.e. 1 meter of sediment deposited on average in the channel over three months).

3.3 *Model validation*

The model was then validated with data from the flushing operation of May-June 2010. Two bathymetric surveys preceded and followed this flushing, in April and August 2010, respectively.

The layer construction is based on values and results of the calibration:

- An old sediment layer is comprised between the non-erodible bottom and the result of the 2008 COURLIS calculation.

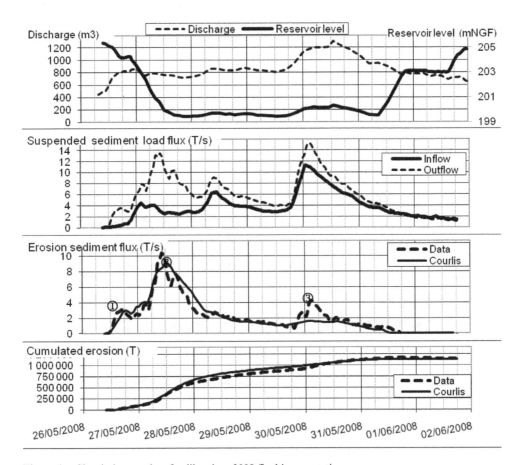

Figure 4. Simulation results of calibration, 2008 flushing operation.

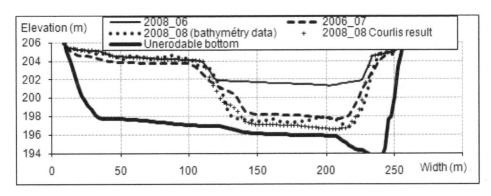

Figure 5. Evolution of a cross-section profile after the 2008 flushing (distance from the reservoir: 1000 m).

- A recent sediment layer is comprised between the bottoms produced by the 2008 COURLIS calculation and the bathymetric profiles of April 2010 translated by −30 cm.
- A layer of slightly consolidated material is constructed from the bathymetric profiles of April 2010 and the same one translated by −30 cm in the channel.

The flushing parameters and the results are summarized in Figure 6.

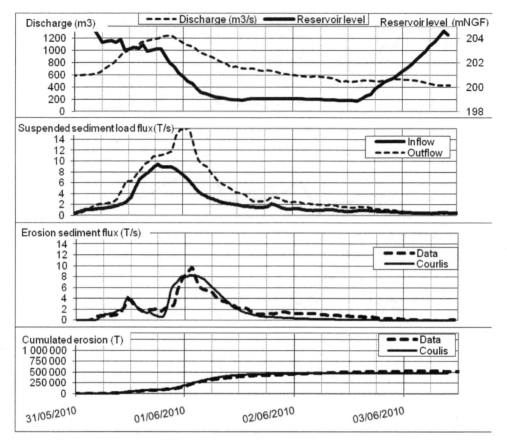

Figure 6.   Validation calculation results, 2010 flushing operation.

The following points should be highlighted:

- During the phase of the lowering of the water level, we clearly see the erosion peak due to the slightly consolidated sediment.
- The main erosion peak, corresponding here to the passage of the flood peak during the lowering phase, is well represented.
- The flux at the end of the episode is underestimated in comparison with measurements.

The correlation between the calculated mass of eroded sediment (0.48 Mt) and the measured one (0.52 Mt) is satisfactory. The evolution of the eroded mass over time is also well represented (the scale used on this graph during the calibration was maintained). The analysis of cross-section profiles (not presented in this paper) yields results similar to those presented for the calibration.

3.4   Influence of parameters: number of layers

The erosion stress, the surface erosion rate and the number of layers are the main parameters of the erosion module and have a strong influence on the results. An illustration of the influence of the number of layers is given in Figures 7 and 8. Assuming only one single layer it is impossible to determine correctly the evolution of the erosion for the considered floods. The use of a single stress of 9 Pa, given acceptable results for the flood 2008, leads to an excessively low value for the 2010 flood.

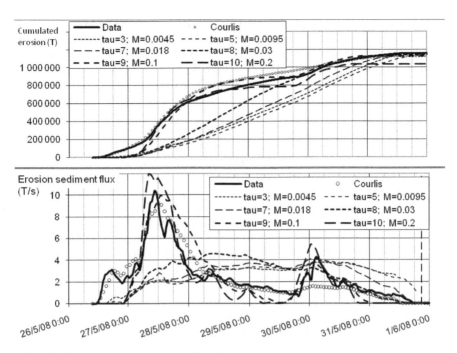

Figure 7.   Single constant layer model—calibration result.

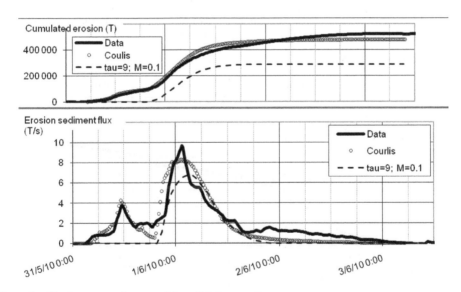

Figure 8.   Single constant layer model—validation results.

4   2D/3D MORPHODYNAMICS SIMULATIONS

Additional, 2D and 3D numerical computations were also performed using (*i*) the 2D hydro-dynamic module TELEMAC-2D, internally coupled to the two-dimensional sediment trans-port module SISYPHE, and (*ii*) the 3D hydrodynamics module TELEMAC-3D, internally coupled to SISYPHE.

Our objective was to compare the different modules of the TELEMAC-MASCARET System in this simple elongated geometrical configuration which is perfectly adapted for 1D simulation, even if the 2D and 3D simulations would give more insight in the detailed structure. This section gives the first simulation results. This is an ongoing work still in progress.

The erosion shear stress parameter τ_{eff} used in SISYPHE is defined as:

$$\tau_{eff} = \tau.\left(\frac{C'_d}{C_d}\right) = \rho g \frac{U^2}{K^2 h^{\frac{1}{3}}}.\left(\frac{C'_d}{C_d}\right) \tag{3}$$

where U = depth-averaged velocity (m/s), C_d et C'_d are both quadratic friction coefficients related to total friction and skin friction, respectively.

Both 2D and 3D hydrodynamic models use the same value of the Strickler friction coefficient (65 m$^{1/3}$ s^{-1}). Both 2D and 3D models are therefore equivalent and give similar results regarding the main hydrodynamics parameter for sediment transport applications (the bed shear stress). The only difference comes from the directions: in 2D, the bed shear stress is supposed to be aligned with the mean flow direction, whereas in 3D, the bed shear stress is aligned with the near bed flow.

4.1 Multi-layer model development

As in COURLIS, bed layers have been defined on the basis of historical bathymetries in SISYPHE, except the unreadable bottom (fixed as a constant level). SISYPHE model parameters are summarized in Table 1.

Table 1. Sedimentary layer setup.

Sediment layers	Concentration (kg/m³)	Critical erosion shear stress (Pa)	Erosion parameter (kg/s/m²)
(a) Slightly consolidated	1000	2	0.004
(b) Recent	1100	7	0.04
(c) Old	1100	9	0.05

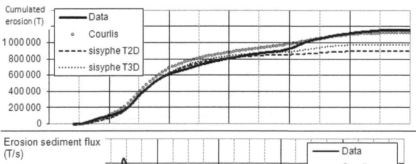

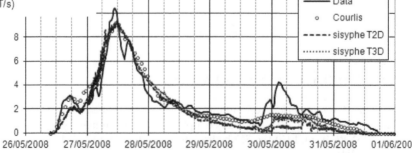

Figure 9. Cumulated erosion and total flux for the different models (2008 flushing).

4.2   Comparison of model results

The comparison between all model results and the data for the net eroded sediment fluxes are given in Figure 9 and 10. For the 2008 flushing event, all models give globally similar results and successfully reproduce the two first peaks (1, 2), but underestimate the third one corresponding to the flood peak.

The cumulated erosion is less well represented with SISYPHE. There's still an ongoing work to improve these results on the basis of:

- a better definition of unerodable bottom,
- an improvement of the calibration parameters
- The use of a model of the reservoir that includes the confluence area.

Another improvement of SISYPHE could concern the definition of the bed shear stress law. As perceptible in Figure 11, SISYPHE over-deepens the bottom in the center of the channel. The same behaviour was obtained by COURLIS by using the equation (1). The choice of the equation (2) in the software improves results as shown on the Figure 5.

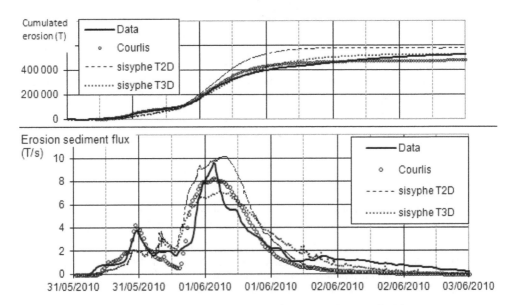

Figure 10.   Cumulated erosion and total flux for the different models (2010 flushing).

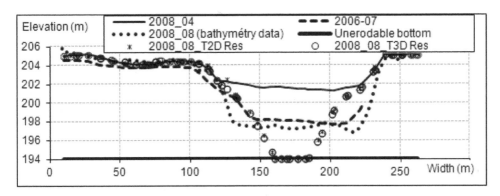

Figure 11.   Evolution of a cross-section profile after the 2008 flushing (dist. from the reservoir: 1000 m).

193

5 CONCLUSION

Three hydrodynamic and sediment transport models of the Telemac-Mascaret System, namely COURLIS/MASCARET (1D) and SISYPHE coupled to TELEMAC-2D and -3D, have been applied to simulate the effect of a flushing event.

The computational domain presented an elongated, unidirectional geometrical configuration, which is particularly well adapted for a 1D application and, therefore, allowed a fair comparison with the 2D and 3D models. We emphasize that the choice of the spatial dimension to which apply the different models depends mainly on the scale of the problem (time and space) and the degree of detail of application.

Currently the cumulated erosion is best fitted by COURLIS. There's still an ongoing work to improve SISYPHE results on the basis of:

- a better definition of unerodedable bottom,
- an improvement of the calibration parameters,
- The use of a model of the reservoir that includes the confluence area.

Another improvement of SISYPHE could concern the definition of the bed shear stress law. As perceptible in Figure 11, SISYPHE over-deepens the bottom in the center of the channel. The same behaviour was obtained by COURLIS by using the equation (1). The choice of the equation (2) in the software improves results as shown on the Figure 5.

Future work will include the implementation, verification and validation of the fully 3D cohesive sediment transport processes within TELEMAC-3D.

For more complex configurations than the one presented here, 1D models can be used to simulate the entire reach, providing then the boundary conditions for more detailed 2D or 3D analysis in important subreaches.

REFERENCES

Bouchard, J.P. & Bertier, C. 2008. Morphological change in reservoirs in relation to hydraulic conditions.

Brandt, S. 2000. Classification of geomorphological effects downstream of dams. *Catena* 40, 375–401.

Jodeau, M. & Menu, S. 2012. Sediment transport modeling of reservoir drawdown, example of Tolla reservoir.

Mahmood, K. 1987. Reservoir sedimentation: impact, extent, and mitigation.

Marot, D., Bouchard J.P. & Alexis A. 2005. Reservoir bank deformation modeling: application to Grangent reservoir. *journal of hydraulic engineering* 131 (7), 586–595.

Morris, G.L. & Fan, J. 1997. Reservoir sedimentation handbook. McGraw-Hill.

Reservoir Sedimentation – Schleiss et al. (Eds)
© 2014 Taylor & Francis Group, London, ISBN 978-1-138-02675-9

Sequential flushing of Verbois and Chancy-Pougny reservoirs (Geneva, Switzerland)

E.F.R. Bollaert
AquaVision Engineering Sàrl, Ecublens, Switzerland

S. Diouf & J.-L. Zanasco
Services Industriels de Genève, Geneva, Switzerland

J. Barras
Société des Forces Motrices de Chancy-Pougny, Chancy, Switzerland

ABSTRACT: Full drawdown flushing has taken place at the Verbois and Chancy-Pougny reservoirs, just downstream of the City and Lake of Geneva, in June 2012. About 700,000 m³ of suspended sediments are transported annually by the Arve River. More than 50% of this material deposits in the Verbois reservoir, the remaining being transported towards downstream.

The 2012 flushing event has evacuated about 2.7 millions of m³ of sediments. Previous flushing dated from 2003, resulting in an exceptionally high volume of sediments being displaced. The 2012 operation revealed to be delicate because of its potential influence on the water level of Lake of Geneva (reduced outflows) and because of its potential environmental impact on the Rhône River downstream.

The paper describes the exceptional sequence of unattended events that occurred, as well as resulting implications for both up-and downstream areas. Bathymetric surveys taken after the event revealed that more than 200,000 m³ of sediments had deposited in the Chancy-Pougny reservoir, increasing the maximum water levels during floods and compromising safety. Detailed 2D numerical modeling of hydraulics and morphology allowed to determine that the major part of these sediments could be displaced towards downstream by applying partial drawdown flushing events of the Chancy-Pougny reservoir. In June 2013, a partial drawdown flushing has taken place during 36 hours.

The paper compares the observed and numerically predicted morphological changes of the Chancy-Pougny reservoir during partial drawdown and points out the excellent predictions that were obtained by the numerical model. It also points out the current investigations to modify the full drawdown flushings at Verbois reservoir into partial drawdown events, in order to reduce environmental impact downstream.

1   INTRODUCTION

Full drawdown flushing has taken place at the Verbois and Chancy-Pougny reservoirs, situated just downstream of the City and the Lake of Geneva, in June 2012 (see Fig. 1). About 700,000 m³ of suspended sediments are transported annually by the Arve River, which has a total catchment area of 2,000 km², including the Mont-Blanc glacier area. More than 50% of this material deposits in the Verbois reservoir, the remaining being transported by the Rhône River, through the Chancy-Pougny reservoir, towards downstream. The Verbois dam is about 34 m high and 340 m long, containing 4 large overflow gates and 4 bottom gates along the left bank, and a powerhouse with 4 Kaplan turbines situated along the right bank.

The 2012 flushing event has allowed displacing about 2.7 millions of m³ of sediments towards downstream. Exceptionally, the event has taken place during 6 days in order to allow

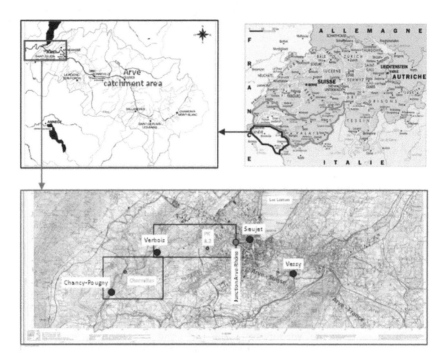

Figure 1. Catchment area of Arve River and localization of Verbois and Chancy-Pougny reservoirs.

replacement of the trash-racks of the water intakes of the power house. Also, previous flushing dated already from 2003, while normally only a 3-year interval is being applied, resulting in an exceptionally high volume of sediments being displaced. For these reasons, the 2012 operation revealed to be particularly delicate because of its potential influence on the water level of Lake of Geneva (reduced outflows during 6 days) and because of its potential environmental impact.

2 THE 2012 FLUSHING EVENT

2.1 Flushing characteristics

The flushing characteristics are illustrated at Figure 2 and the sediments are visualized at Figure 3. During a first stage, the normal reservoir level is being drawdown from 369 m a.s.l. down to 354.7 m a.s.l. during 60 hours. This drawdown is being performed at different speeds, for outflows between 200 and 400 m³/s, to control the max. suspended sediment concentration of the water leaving the reservoir. Concentrations of 30–35 g/l were expected, based on computations.

Next, the reservoir level is being held low at 354.7 m a.s.l. to perform flushing of sediments and to allow a replacement of the trashracks of the water intakes towards the turbines. The former needs high outflows of about 500–600 m³/s during 2 days, while the latter had never been done before and was expected to last about 6 days in total, for a low discharge of about 200 m³/s. The total drawdown duration has thus been largely extended compared to a normal flushing event. Finally, a quite rapid and constant filling up of the reservoir is planned.

2.2 Unattended events during flushing

The flushing event started on June 9th 2012 at midnight, referred to as 0 h in Figure 4. Figure 4 also illustrates the most important unattended events as described in detail hereafter.

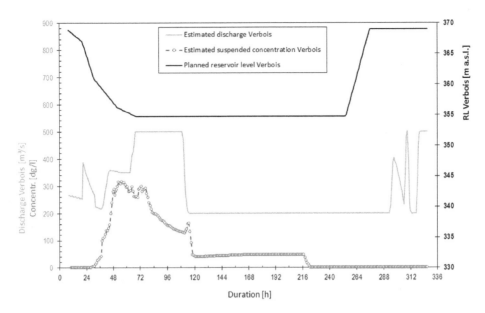

Figure 2. Estimated 2012 flushing characteristics of the Verbois reservoir.

Figure 3. Photo of Verbois reservoir during complete drawdown (looking towards upstream).

2.2.1 *Heavy rainfall and discharges in France*

About 24h after start of flushing, heavy rainfall on both the Arve and Rhône catchment areas in France resulted in discharges of the downstream Rhône River that were much higher than predicted. As such, during the first few days of the flushing event (up to 100h at Fig. 4), and to satisfy French requirements, the outflow at Verbois has been voluntary reduced, in order to limit as much as possible the total Rhône discharge along its French course down to Lyon.

While the expected sediment concentrations were not significantly affected by the outflow reduction, the total volume of sediment evacuated from the reservoir was much smaller than expected. Hence, to compensate this lack of flushing efficiency, outflows of up to 500 m³/s have been maintained as much as possible during the next 2 days (i.e. 120h–168h in Fig. 4).

2.2.2 *Flow instabilities at the bottom outlets*

The bottom gates were expected to maintain the reservoir level at 354.7 m a.s.l. This value was the lowest possible value that still avoids inundation of the workers area where the intake

197

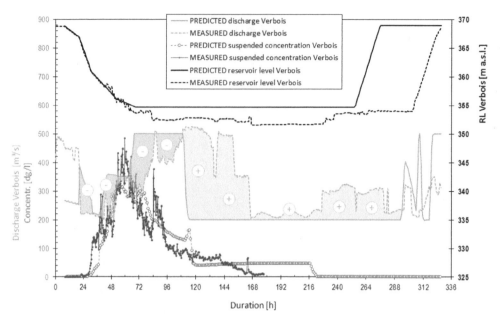

Figure 4.   MEASURED 2012 flushing characteristics of the Verbois reservoir (Diouf, 2012).

trashracks were being replaced, but still high enough such to minimize peak suspended concentration towards downstream (environmental concerns).

Nevertheless, severe flow instabilities at the bottom gates asked for a lower water level to avoid inundation of the workers area. For most of the flushing, a value of about 353 m a.s.l. has been adopted, which resulted in an increase of the flushed sediment volume.

2.2.3   *Landslide of mud into the turbine intakes*

After about 48h of flushing, due to the lowering of the reservoir level, a landslide occurred next to the turbine intakes. This event resulted in 5,900 m³ of mud covering both the workers area and the 4 turbine intakes. The mud was found all inside the intakes. A total of 7 machines worked during 72 h non-stop to clean the area and allow to continue the trashrack replacement. The event retarded the flushing event by about 48 h.

2.2.4   *Critical rise of Lake of Geneva water level*

At the same time, the period of heavy rainfall resulted in a significant rise of the water level of Lake of Geneva, due to very high inflows from Wallis. Figure 5 illustrates the evolution of the lake level during the flushing event at Verbois. While the normal max. lake level is defined at 372.30 m a.s.l., a level of 372.50 m a.s.l. was accepted during the event, in accordance with the different Cantonal Services, as the level not to exceed. Indeed, above this level, several locations around the lake start to be inundated.

Lake level predictions made on the 13th of June indicated that a max. level above 372.65 m a.s.l. would be reached near the end of the flushing and thus action had to be taken. Fortunately, this action was in agreement with the necessary increase of discharge at Verbois, needed at that time to enhance flushing efficiency (Fig. 4, 120h–168h). Second, lake level predictions on the 18th of June again revealed that the critical level would be reached near the end of the flushing. Hence, during 48 h, the outflow at Verbois was increased from 200 to 300 m³/s.

This increase in discharge was a very delicate operation, because the total volume of evacuated sediments was already reached and the Chancy-Pougny reservoir just downstream was already filled up at that time. In other words, outflow of sediments had to be absolutely minimized to avoid deposition of sediments in the Chancy-Pougny reservoir.

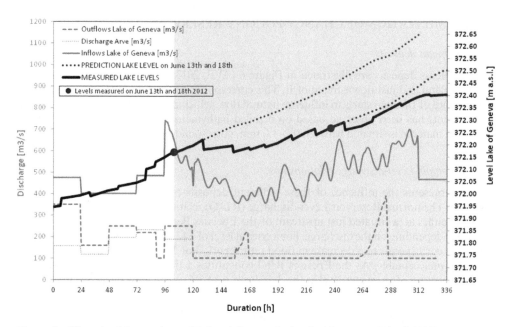

Figure 5.   Water level fluctuations of Lake of Geneva during flushing event (Diouf, 2012).

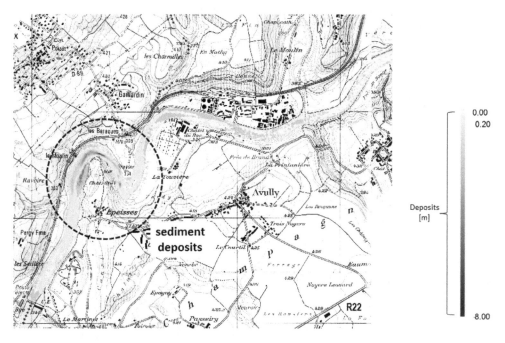

Figure 6.   Sediment deposits in the Chancy-Pougny reservoir following the 2012 flushing event.

2.2.5 *Excessive sediment deposits in the Chancy-Pougny reservoir*
Unfortunately, this was exactly what happened. About 210,000 m³ of sediments deposited in the Epeisses Bend of the reservoir (Fig. 6). As Chancy-Pougny is a run-of-river reservoir, this resulted in a significant increase of the reservoir levels for a given discharge and thus the risk of inundations in upstream areas during high flood events.

3 SEDIMENTATION OF CHANCY-POUGNY RESERVOIR

3.1 Sediment deposits

The sediments deposits are illustrated in Figure 6 (AVE, 2013). They are concentrated at the Epeisses Bend and just downstream of it. This corresponds to the area of the reservoir where the presence of the dam starts to influence natural flow velocities. A total volume of 210,000 m³ of sediments has been observed based on detailed bathymetric surveys. The diameter of the deposited material is situated between 0.2–1.0 mm, corresponding to fine to mean sand.

3.2 Reservoir Level-discharge relationships

Figure 7 presents the influence of these sediment deposits on the backwater curves of the reservoir. The normal Reservoir Level-discharge (RL-Q) relationship, maintained at the Nant des Charmilles area situated just upstream of the Epeisses Bend, is shifted upwards by about 0.6–1.3 m depending on the discharge. For a critical level of 347.95 m a.s.l. at which inundations start, this corresponds to a decrease in flood return period of 100 years to only 5 years, which is clearly unacceptable. As the Epeisses Bend constitutes a flow control zone for upstream, further lowering of the water level at the dam itself is of no use to solve the problem.

3.3 Numerical optimization of potential countermeasures

As such, different countermeasures have been numerically tested and optimized:

1. Dredging of sediments in the most critical areas of the Epeisses Bend
2. Flushing of Chancy-Pougny reservoir by partial drawdown during future flood events

As shown in Figure 8, dredging of a volume of 100,000 m³ of sediments in the most critical areas solves half of the problem but does not allow getting back to normal reservoir operational levels. Given the relatively high costs of dredging, this did not seem to be a plausible solution.

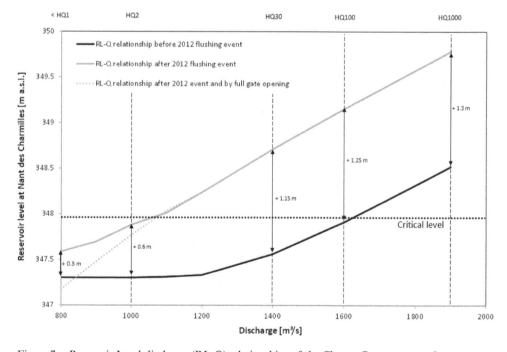

Figure 7.   Reservoir Level-discharge (RL-Q) relationships of the Chancy-Pougny reservoir.

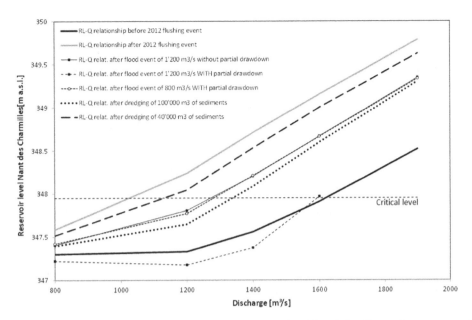

Figure 8. Computed Reservoir Level-discharge (RL-Q) relationships of the Chancy-Pougny reservoir in case of dredging or partial drawdowns during future flood events (AVE, 2013).

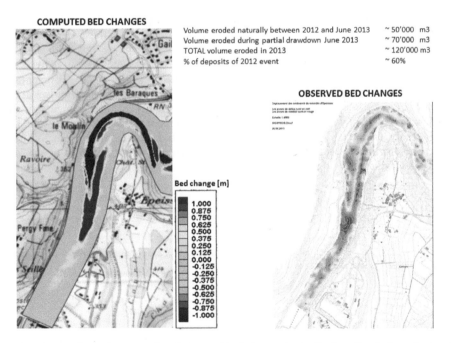

Figure 9. Comparison of computed and observed bed changes in the Epeisses Bend after a first partial drawdown of −1.5 m at the dam for an average discharge of 800 m³/s.

Second, based on 2D morphological computations, a significant part of the sediments may be transported towards downstream during future natural or artificially generated flood events. This phenomenon is clearly accentuated by a partial drawdown of the normal operating level at the dam itself (Fig. 8). As such, a drawdown of about 2 m at the dam, for a flood event of 1,200 m³/s, would even allow obtaining an operating level lower than the official one.

Hence, the latter solution has been preferred. In total, 3 partial drawdowns have been planned, for discharges of 800–1,000 m³/s and with partial drawdowns at the dam of at least 1.0–1.5 m. The general idea is to learn from the first drawdown operation, in order to optimize the subsequent ones, if needed.

3.4 *Partial flushing of the Chancy-Pougny reservoir on 21–22 June 2013*

Suitable hydraulic conditions were met on 21st–22nd June 2013. For an average discharge of 800 m³/s, and a drawdown of 1.5 m at the dam, about 70,000 m³ of sediments were eroded from the Epeisses Bend and re-deposited further downstream in the reservoir. This erosion and re-deposition occurred exactly as predicted by 2D numerical computations, as shown in Figure 9.

Furthermore, between the 2012 Verbois flushing and the 2013 Chancy-Pougny partial drawdown, about 50,000 m³ of very fine sediments were already transported out of the reservoir during natural flood events. As such, a total of 120,000 m³ of sediments, i.e. about 60% of the total volume, has been displaced by the end of 2013.

Hence, the new operating levels of the reservoir correspond to the curve valid for 100,000 m³ of dredging (Fig. 8). As such, about 2/3 of the level increase has been compensated.

Given the success of this operation and based on the numerical predictions that revealed to be in excellent agreement with the observations, more partial drawdowns are planned in the future to bring the operating levels completely back to their normal values.

4 CONCLUSIONS

The 2012 Verbois flushing evacuated 2.7 millions of m³ of sediments, from which more than 200,000 m³ deposited in the Chancy-Pougny reservoir, increasing the water levels during floods and compromising safety. Detailed 2D numerical modeling of hydraulics and morphology allowed to optimize partial drawdown flushing of the Chancy-Pougny reservoir to bring the water levels back to their normal values. Due to significant environmental concerns along the Rhône River in both France and Switzerland, plausible alternatives are being studied for the Verbois flushing. Current investigations focus on the pertinence of partial drawdowns combined with measures that facilitate sediment transfer during natural floods. These are, for example, dilution of the Arve sediments by clean water from the Lake of Geneva, or also increase of sediment transport capacity by minor lowering of the Verbois reservoir level during floods.

REFERENCES

AVE. 2013. *Internal Report*, Services Industriels de Genève, Suisse.
Diouf, S. 2012. Vidange de Verbois: Tome I: Bilan hydraulique et sédimentaire, *Internal Report*, Services Industriels de Genève, Suisse.

Reservoir Sedimentation – Schleiss et al. (Eds)
© 2014 Taylor & Francis Group, London, ISBN 978-1-138-02675-9

Simulation of the flushing into the dam-reservoir Paute-Cardenillo

Luis G. Castillo & José M. Carrillo
Hidr@m Group, Universidad Politécnica de Cartagena, Cartagena, Spain

Manuel A. Álvarez
GEAMA Group, Universidade da Coruña, Coruña, Spain

ABSTRACT: The study analyzes the expected changes in the Paute River in Ecuador as a result of the construction of the Paute-Cardenillo Dam (owned by Celec Ep-Hidropaute). This dam will integrate the National Electric System of Ecuador. Given that the project must remain viable throughout its useful life, the operational rules at the reservoir are required to include sedimentation effects. Sediment transport and flushing are studied by using four complementary procedures: empirical formulae, one-dimensional simulations (time required for sediment level to reach the height of the bottom outlets), two-dimensional simulations (flushing) and three-dimensional simulations (detail of the sediment transport through bottom outlet). Besides this, three-dimensional simulations have been used to considerer the effect of increasing the roughness due to the sediment transport in the bottom outlets.

1 MAIN CHARACTERISTICS OF THE PROJECT

The study zone is situated in the Paute River basin in Ecuador 23 km downstream from the Amaluza Dam. The area to be analyzed is of 275 km² of draining surface and the average slope of the river reach is 0.05 (Fig. 1).

Paute-Cardenillo is a double curvature arch dam with a maximum height of 135 m to the foundations. The top level is located at 926 meters. The reservoir has a length of 2.98 km, with normal maximum water level being located at 924 meters.

Figure 2 shows the sieve curves obtained at three sites of the river and the mean curve used in the calculations.

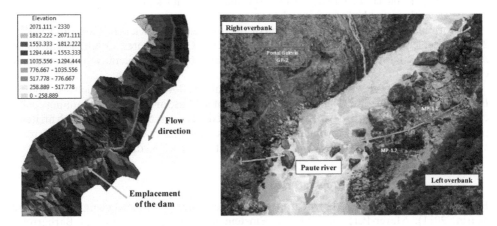

Figure 1.   (a) Zone of study in Paute River basin. (b) Paute river.

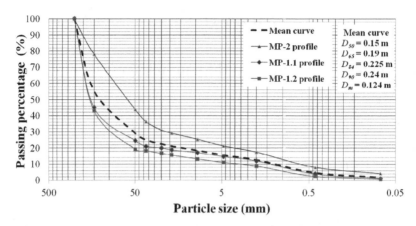

Figure 2. Sieve curves of three sites of the river and the sieve mean curve.

The total bed load (excluding wash load) was determined as 1.75 Mm³/year and the maximum volume of the reservoir 12.33 Mm³. In order to prevent the deposition of sediments into the reservoir, periodic discharges of bottom outlet or "flushing" have been proposed. These operations should be able to remove the sediments, avoiding the advance of the delta from the tail of the reservoir. The hydraulic flushing is considered an efficient technique in narrow reservoirs and strong slopes (Lai and Shen 1996, Janssen 1999).

Initial studies indicate that the minimum flow evacuated by the bottom outlet to achieve an efficient flushing should be at least twice the annual mean flow ($Q_{ma} = 136.3$ m³/s). For the safe side, a flow of 408.9 m³/s ($3Q_{ma}$) was adopted.

2 SEDIMENT TRANSPORT FORMULAE

Sediment transport may be divided into the following: wash load (very fine material transported in suspension), and total bed transport (bed sediment transported and/or in suspension, depending on the sediment size and flow velocity). The main properties of sediment and its transport are: the particle size, shape, density, settling velocity, porosity and concentration.

2.1 Estimation of the manning resistance coefficient

Following the methodology applied in Castillo et al. (2000), four aspects were checked to determine hydraulic characteristics of the flow: macro roughness, bed form resistance, hyper concentrated flow, and bed armoring phenomenon.

Ten formulae were applied for estimation of the roughness coefficient (Castillo and Marin, 2011): Strickler, Limerinos, Jarret, Bathurst, van Rijn, García-Flores, Grant, Fuentes and Aguirre-Pe, and Bathurst. These formulae are calculated by coupling iteratively the hydraulic characteristics with the sediment transport.

A macro roughness behavior may be identified in all the flows analyzed, which also present the armoring phenomenon. This leads to a significant increase in the various Manning coefficients. The calculation of these coefficients was carried out in a section type, through an iterative procedure by using the formulation of Fuentes and Aguirre-Pe (1991) and Aguirre-Pe et al. (2000). Figure 3 shows the Manning coefficients for the flow rates.

2.2 Estimation of sediment transport

Fourteen formulations of sediment transport capacity were used coefficient (Castillo and Marin, 2011): Meyer-Peter and Müller, Einstein and Brown, Einstein and Barbarrosa, Colby, Engelund and Hansen, Yang, C.T., Parker et al., Smart and Jaeggi, Mizuyama and

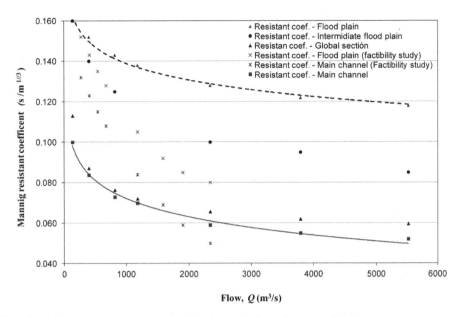

Figure 3.   Manning resistance coefficients in the main channel and floodplain.

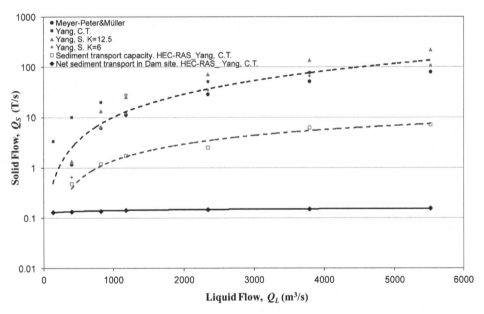

Figure 4.   Sediment transport capacity (reach mean values), sediment transport (simulation of all reach) and net sediment transport in dam site.

Shimohigashi, van Rijn, Bathurst et al., Ackers and White, Aguirre-Pe et al., and Yang, S. From these, the formulations that fell within a range of the mean value +/− 1standard deviation were selected. Figure 4 indicates that the transport capacity could vary between 1 and 100 t/s, if the mean values of the analyzed reach are considered. However, these values are reduced between 0.5 and 10 t/s when the river complete reach is considered (erosion and sedimentation processes are simulated). Finally, the net sediment transport in dam site was only 0.2 t/s.

3 NUMERICAL SIMULATIONS

The bed level change Z_b can be calculated from the overall mass balance equation for bed load sediment (Exner equation):

$$(1-p)\frac{\partial Z_b}{\partial t} + \frac{\partial Q_{bs}}{\partial s} + \frac{\partial Q_{bn}}{\partial n} = 0 \tag{1}$$

where p is porosity of the bed material; Q_{bs} and Q_{bn} are the bed load flux in the main flow direction s and in the cross flow direction n, respectively. They are calculated from the non-equilibrium bed load equation:

$$\frac{\partial(Q_b\alpha_{bs})}{\partial s} + \frac{\partial(Q_b\alpha_{bn})}{\partial n} = -\frac{1}{L_s}(Q_b - Q_e) \tag{2}$$

where α_{bs} and α_{bn} are the cosines of the direction vector that determines the components of the bed load transport in the s and n directions, respectively. The model considers that the non-equilibrium effects are proportional to the difference between non-equilibrium bed load Q_b and equilibrium bed load Q_e, and related to the non-equilibrium adaptation length L_s.

3.1 Reservoir sedimentation

The time required for sediment level to reach the height of the bottom outlets (elevation 827 m) operating at reservoir levels was analyzed. Simulations were carried out with the one-dimensional HEC-RAS 4.1 program which employs a continuity equation of sediment.

The input flows are the annual mean flow ($Q_{ma} = 136.3$ m³/s) equally distributed in the first 12 km and the incorporation (2.44 km upstream from the Paute-Cardenillo Dam) of the annual mean discharge flow of the Sopladora hydroelectric power plant ($Q_{ma\_sop} = 209.0$ m³/s).

The suspended sediment concentration at the inlet section was 0.258 kg/m³. This value is similar to the mean concentration at the Sopladora hydroelectric power plant. The sediment characteristic diameter in the dam emplacement was $D_{50} = 0.150$ m. The sediment transport was calculated by considering the Meyer-Peter and Müller (1948), and the Yang (1976), formulae.

Table 1 shows the volume of sediments in the reservoir obtained when the bottom outlet level was reached. According to the results, the volume of sediment in the reservoir rises with the increasing of the water level in the reservoir, and requires a longer duration to reach the bottom outlet elevation.

The least favorable condition (the first one in which the sediment reaches the elevation of 827 m) was obtained with the expression of Meyer-Peter and Müller and the level of the reservoir located at 860 m, requiring a time of 3 months and 27 days.

3.2 Flushing simulation

3.2.1 Two-dimensional simulation

The flushing process was analyzed by using the Iber two-dimensional program. Iber can be divided in three modules: hydrodynamic, turbulence and sediment transport. The program

Table 1. Time required and volume of sediment when the bottom outlets are reached.

Reservoir elevation	Yang		Meyer-Peter & Müller	
	Required time (years)	Sediment volume (hm³)	Required time (years)	Sediment volume (hm³)
860	0.33	0.65	0.33	1.47
918	12.90	6.07	8.80	7.34

uses triangular or quadrilateral elements in an unstructured mesh and finite volume scheme. The hydrodynamic module solves shallow water equations (2-D Saint-Venant equations). Diverse turbulence models with various levels of complexity can be used. The sediment transport module solves the transport equations by the Meyer-Peter and Müller expression and the evolution of the bottom elevation is calculated by sediment mass balance.

According to the operational rules of the Paute-Cardenillo Dam, the evolution of the flushing is studied over a continuous period of 72 hours. The initial condition of sedimentation profile was obtained with the HEC-RAS simulation (1.47 hm³ of sediment). The input flow was three times the annual average flow Wan and Wang (1994) (408.9 m³/s). The suspended sediment concentration at the inlet section was 0.258 kg/m³. The initial water level at the reservoir was 860 m. Effective flushing was observed during the operation of the bottom outlets. Figure 5 shows the profiles of the sediment at the reservoir during the flushing operation.

After a flushing period of 72 hours, the volume of sediments removed by the bottom outlets was 1.76 hm³. This value is bigger than the initial sediment value (1.47 hm³).

3.2.2 Three-dimensional simulation

The Computational Fluid Dynamics (CFD) program FLOW-3D, which solves the Navier-Stokes equations discretized by finite differences, was used. It incorporates various turbulence models, a sediment transport model and an empirical model bed erosion (Guo, 2002; Mastbergen and Von den Berg, 2003; Brethour and Burnham, 2010), together with a method for calculating the free surface of the fluid without solving the air component (Hirt and Nichols, 1981). The bed load transport is calculated by using the Meyer-Peter & Müller (1948) and Van Rijn (1986) expressions.

The operation of the bottom outlets, starting from the initial conditions of sedimentation obtained with the HEC-RAS program, was analyzed. Due to the high concentration of sediments that pass through the bottom outlets, the variation of the roughness in the bottoms has been considered. The estimation of the hyperconcentrated flow resistance coefficient on rigid bed has been calculated by using the formulae proposed by Nalluri (1992):

$$\lambda_s = 0.851\lambda_0^{0.86} C_v^{0.04} D_{gr}^{0.03} \tag{3}$$

where λ_s = Darcy-Weisbach's resistance factor on rigid bed with sediment transport, λ_0 = Darcy-Weisbach's resistance factor on rigid bed with clean water, C_v = volumetric sediment concentration, D_{gr} = grain size non-dimensional factor.

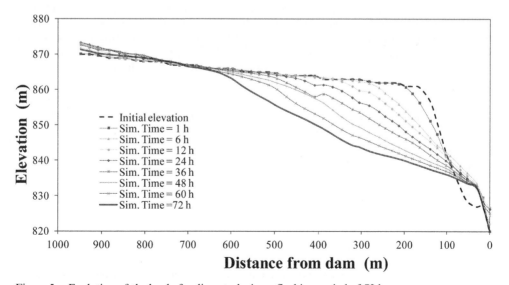

Figure 5.   Evolution of the level of sediments during a flushing period of 72 hours.

The bottom outlets are four rectangular ducts (5.00 × 6.80 m). The slope corresponding to the stretch under study is $S = 0.001$. The C_v has been estimated in 0.04. According to Wan and Wang (1994), the energy supplied by the solid phase, for a volume unit and a distance unit downstream (in non-dimensional way) is:

$$\frac{E_d}{\gamma} = C_v S = 0.00014 < 0.004 \tag{4}$$

This value is really lower than the limit between the hyperconcentrated flow and mudflow (0.004). The coefficient of cinematic viscosity of water with sediments concentration has been estimated by using the following formulae (Graf, 1984):

$$\frac{v_s}{v} = 1 + K_e C_v + K_2 C_v^2 \tag{5}$$

where v = the coefficient of cinematic viscosity of clean water (for T = 15°C, $v \cong 1.14 \times 10^{-6}$ m²/s), K_e = Einstein viscosity constant ($\cong 2.5$), K_2 = particles interaction coefficient ($\cong 2$). Therefore, $v_s = 1.26 \times 10^{-6}$ m²/s, a value 10.5% higher than v. The non-dimensional size of the grain corresponding to $D_{50} = 0.150$ m is:

$$D_{gr} = D_{50} \left(\frac{(s-1)g}{v_s^2} \right)^{1/3} = 3254.74 \tag{6}$$

Ducts are covered by iron, so the absolute roughness $\varepsilon = 5 \times 10^{-5}$ m. With the Coolebrok-White formula, $\lambda_0 = 8.32 \times 10^{-3}$ has been obtained. By using the Nalluri formula, $\lambda_s = 1.55 \times 10^{-2}$. This value corresponds to an absolute roughness $\varepsilon_s = 2.37 \times 10^{-3}$ m. The value ε_s has been used in the three-dimensional simulations.

Due to the high-capacity equipment and long simulation times to calculate the flushing of all the reservoir during 72 h, the results are focused in the first 10 h. The initial conditions of sedimentation was obtained with the HEC-RAS program. The inlet boundary was situated upstream the dam reservoir, considering that the water level is 860 m in the start condition.

Figure 6 shows the discharge flow at each bottom outlet, together with the total flow discharged for the first 3000s of simulation, considering the roughness of the bottom outlets according to clean water and sediments transport. Outlets worked in a pressured and unsteady regime at the initial emptying of the reservoir, reaching a discharge near 2700 m³/s. After the steady regime was reached (around 1000 s of simulation), there was a free surface flow and the discharged flow was the expected (408.90 m³/s during the flushing operation). With the smooth roughness, the bottom outlets are block due to the high sediment transport (approximately 200 s after the beginning), although later it is swept without problems.

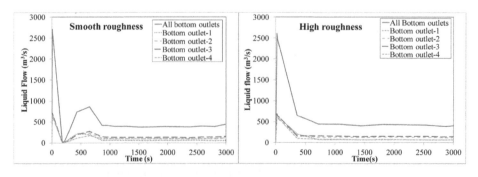

Figure 6. Liquid flows considering smooth and high roughness during the flushing operation simulated with three-dimensional program.

Figure 7 shows the solid flow at each bottom outlet, together with the total solid flow discharged for the first 3000 s of simulation. The roughness of the bottom outlets has been considered with clean water and sediments transport. With the smooth roughness, the peak of sediment transport (430,000 kg/s) is bigger than with the high roughness (310,000 kg/s). The peaks appear during the first steps of the flushing due to the emptying of the reservoir. Later, simulations seem to reach a constant rate of sediments removed, which is higher in the case of ducts with smooth roughness.

Figure 8 shows the volume of sediment removed and the transient sediment transport during the first ten hours of operation. Like in the two-dimensional simulation, there is significant sediment transport at the beginning of the simulation. There is a maximum of 160 m^3/s of sediments near the first hour with the smooth roughness, while it is reduced to 120 m^3/s for the high roughness. Later, the sediment transport rate in both cases tends to decrease until near 6 m^3/s which is similar to the two-dimensional result. The total volume of sediment calculated by FLOW-3D is much higher than with Iber program due to the simulations of the flushing phenomenon are very different in the first three hours.

The two-dimensional simulations considered that all the sediments (1.47 hm^3) may be removed in 60 hours. Considering that the rate of sediments is going to continue during the rest of the flushing operation, the three-dimensional simulations with the high roughness would require near 45 hours to remove all the sediments. Hence, the Iber simulations are from the safety side. They require longer times to remove the same volume of sediment than the three-dimensional simulations (theoretically more accurate).

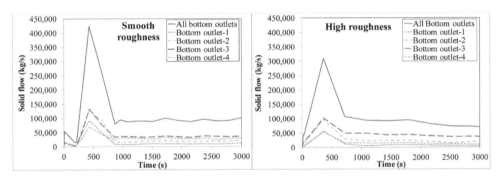

Figure 7.   Solid flows considering smooth and high roughness during the flushing operation simulated with three-dimensional program.

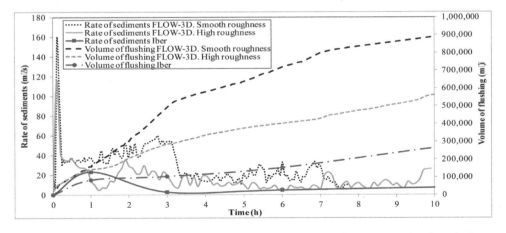

Figure 8.   Comparison of the flushing operation simulated with two-dimensional and three-dimensional simulations.

4 CONCLUSIONS

In this paper, the complex phenomenon of flushing has been analyzed by using four interrelated methodologies: empirical formulations, one-dimensional simulations, two-dimensional simulations and three-dimensional simulations. Empirical simulations constitute an upper envelope of the sediment transport capacity. This procedure allowed an estimate of the coefficients of resistance or Manning roughness applied in the numerical simulations.

Due to the time period (one year) required to analyzing the sedimentation process in the reservoir, and the length of the reach (23.128 km), simulations were carried out with a one-dimensional program (two and three-dimensional programs need high-capacity equipment and long simulation times). Flushing operation was simulated with two-dimensional (Iber) and three-dimensional (FLOW-3D) programs. For 72 hours of flushing simulation, the Iber program required near 24 hours (Intel Core i7 CPU, 3.40 GHz processor, 16 GB RAM and 8 cores). The FLOW-3D program, by using the same equipment, would require more than 2400 h (100 days). Hence, the three-dimensional simulations were only used to analyze the behavior of the flow in the first 10 h of the flushing (343 h of real time each simulation).

The increase of the roughness in the ducts drives to a reduction of the amount of sediment removed in the reservoir.

The results demonstrated the suitability of crossing different methodologies to achieve an adequate resolution of complex phenomena such as flushing operations.

ACKNOWLEDGMENTS

The authors are grateful to CELEC EP—Hidropaute and the Consorcio POYRY-Caminosca Asociados for the data provided.

REFERENCES

Aguirre-Pe, J., Olivero, M.I. and Moncada, A.T. 2000. Transporte de sedimentos en cauces de alta montaña. *Revista Ingeniería del Agua*, 7(4), 353–365 (in Spanish).
Brethour, J. & Burnham, J. 2010. Modeling Sediment Erosion and Deposition with the FLOW-3D Sedimentation & Scour Model. *Flow Science Technical Note. FSI-10-TN85*: 1–22.
Castillo, L. & Marín, M.D. 2011. Characterization of Ephemeral Rivers. *34th IAHR World Congress. Brisbane, Australia.*
Castillo, L., Santos, F., Ojeda, J., Calderón, P., and Medina, J.M. 2000. Importancia del muestreo y limitaciones de las formulaciones existentes en el cálculo del transporte de sedimentos. *XIX Congreso Latinoamericano de Hidráulica. Córdoba, Argentina.*
Fuentes, R. & Aguirre-Pe, J. 1991. Resistance to flow in steep rough streams. *J. Hydraulic Engineering* 116(11): 1374–1387.
Graf, W.H. 1984. *Hydraulics of Sediment transport.* Water Resources Publications, LLC. Colorado, USA.
Guo, J. 2002. Hunter Rouse and Shields diagram, *1th IAHR-APD Congress, Singapore* 2: 1069–1098.
Hirt, C.W. & Nichols, B.D. 1981. Volume of Fluid (VOF) Method for the Dynamics of Free Boundaries. *Journal of Computational Physics* 39(201).
Janssen, R. 1999. *An Experimental Investigation of Flushing Channel Formation during Reservoir Drawdown.* Dissertation presented to the University of California at Berkeley, in partial fulfillment for the requirements for the degree of Doctor of Philosophy.
Lai, J.S. & Shen, H.W. 1996. Flushing sediment through reservoirs. *J. Hydraulic Research* 24(2).
Mastbergen, D.R. & Von den Berg J.H. 2003. Breaching in fine sands and the generation of sustained turbidity currents in submarine canyons. *Sedimentology* 50: 625–637.
Meyer-Peter, E. & Müller, R. 1948. Formulations of the Bed-load Transport. *Proc. of the II IAHR, Stockholm*: 39–64.
Nalluri Chandra 1992. Extended data on sediment transport in rigid bed rectangular channels. *Journal of Hydraulic Research* 30(6): 851–856.
Van Rijn, L.C. 1986. Manual sediment transport measurements in rivers, estuaries and coastal seas. *Rijkswaterstaat and Aqua publications.*
Yang, C.T. 1996. *Sediment Transport: Theory and Practice.* McGraw-Hill International Ed., NY, USA.
Wan, Z. & Wang, Z. 1994. *Hyperconcentrated Flow.* IAHR Monograph Series. Rotterdam, The Netherlands.

Reservoir Sedimentation – Schleiss et al. (Eds)
© *2014 Taylor & Francis Group, London, ISBN 978-1-138-02675-9*

Comprehensive numerical simulations of sediment transport and flushing of a Peruvian reservoir

A. Amini & P. Heller
e-dric.ch Ltd., Switzerland

G. De Cesare & A.J. Schleiss
Ecole Polytechnique Fédérale de Lausanne, Lausanne, Switzerland

ABSTRACT: Numerical modeling of sediment transport in reservoirs, especially for sudden events such as flushing, is still a challenging research topic. In the present study, sediment transport and sediment flushing are simulated for a Peruvian reservoir. Situated in the Peruvian Andes, the watershed is affected by high erosion rates and the river carries high amounts of suspended sediment whose estimated annual volume is about 5 Million m³. This study aims to investigate the reservoir sedimentation using different numerical models. A one-dimensional (1D) sediment transport model, a horizontal two-dimensional (2D) hydraulic model and a vertical 2D model are used for this purpose. Annual sedimentation, full drawdown flushing, and sediment concentration in power intakes are particularly investigated.

1   INTRODUCTION

The present study aims to investigate the reservoir sedimentation aspects by means of numerical simulations. The main purpose is to provide a state of knowledge that can be used later for reservoir sedimentation management. Atkins (1996) developed a technical model for flushing which quantifies aspects of reservoirs that are likely to be successful in flushing at complete drawdown. However, numerical simulations have been used to assess reservoir sedimentation for the last two decades.

Olsen (1999) reproduced the main features of the erosion pattern using a two-dimensional numerical model simulating flushing of sediments from water reservoirs that solves the depth-averaged Navier-Stokes equations on a two-dimensional grid. A 2D horizontal model was used by Bessenasse et al. (2003) for a reservoir in Algeria in which sediment was modeled by concentration thanks to advection-diffusion modelling. Harb et al. (2012) presented the application of **TELEMAC-2D** numerical model for an Alpine reservoir in Austria. Leite et al. (2005) and Möller et al. (2011) applied a three-dimensional numerical model (FLOW-3D) to study pressurized sediment flushing and respectively sediment management in reservoirs.

2   PROJECT AREA AND SPECIFICATIONS

The studied dam is situated in the Peruvian Andes, where the watershed of the project is located in the eastern side of the Andes. Therefore, the flowing water released after several hundreds of kilometers in the Amazonian area is affected by high erosion rates and carries high amounts of suspended sediment.

The hydrology and sedimentology of the catchment need to be fully understood in the planning of flushing facilities for new or existing reservoirs and to provide the background for analyses of past sedimentation and flushing performance. As such, a brief list of the most important parameters is presented hereafter.

2.1 Catchment data

The watershed area of the reservoir is 28'096 km². The reservoir length is 10.2 km with an average river slope of 0.7% at the dam site. A brief list of the most important catchment data is presented in Table 1.

Based on the catchment data and reservoir capacity presented in Table 1 and according to the World Bank experience (Palimieri et al., 2003) for the studied reservoir, where the storage capacity is about 0.5% of the mean annual river run-off, the most adapted remedial measure for this project is regular full drawdown flushing. White (2001) also mentioned that flushing is vital for the preservation of long-term storage where the sediment deposition potential is greater that 1–2% annually of the original capacity, which is the case in the present study.

2.2 Hydrological and sediment data

Daily flow data are available over 46 years with important gaps. The annual mean daily flow is 270 m³/s. The mean daily flow over one year is shown in Figure 1.

In addition, the concentration of suspended load is defined as a function of discharge for each month of the year. The relationship $(Q_s = a \cdot e^b \cdot Q_w)$ gives the suspended load concentration, Q_s, as a function of water discharge, Q_w, where a and b parameters for each month are given. The sediment data for both bed load and suspended load is presented in Table 2.

Table 1.   Watershed data.

Reservoir capacity [million m³]	37
Annual flow [million m³]	7900
Total annual sediment yield [million m³]	5.0
Annual bed load yield [million m³]	1.7
Annual suspended load yield [million m³]	3.3

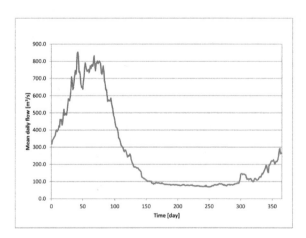

Figure 1.   Mean daily flow.

Table 2.   Sediment data.

Diameter	Bed load	Suspended load
D_{50} [mm]	24	0.05
D_{90} [mm]	87	0.09

212

Table 3.   Geometric data of the dam.

Dam bed level elevation [m asl]	1480.0
Bottom outlet sill level [m asl]	1495.0
Power intake sill level [m asl]	1526.0
Normal water elevation [m asl]	1556.0
Dam crest elevation [m asl]	1560.1
Bottom outlet number [m asl]	6
Bottom outlet dimensions [m]	Height 6.0; width 4.6

2.3   *Dam geometry data*

For effective full drawdown flushing, the bottom outlets must be low enough and of suffi-cient capacity to allow a natural flow through the dam. Some geometrical properties of the dam are summarized in Table 3.

3   METHODOLOGY

To understand and evaluate the reservoir sediment management, several numerical models have been used. A one dimensional model (HEC-RAS) is firstly used to define the bed load and suspended load transport over long time periods up to 5 years. In a second step, a two dimensional horizontal hydraulic model (BASEMENT, with averaged values over the depth) is applied for short term simulations of flushing events. By calculating the hydraulic capacity of the bottom outlets, the bed-level shear stress and consequently flushing efficiency is com-puted. The horizontal 2D model also helps to assure the lateral flow homogeneity and justify the use of a two dimensional vertical model. Finally, the vertical two-dimensional model (CE-QUAL-W2) provides the suspended sediment concentration evolution in the reservoir and brings up the suspended sediment concentration in water intakes for the power plant.

4   ANNUAL SEDIMENTATION

The well-known HEC-RAS model is used to perform a mobile bed sediment transport analysis in the entire reservoir for different long term scenarios. Current sediment capabilities in HEC-RAS are based on a quasi unsteady hydraulic model. The quasi-unsteady approach approxi-mates a flow hydrograph by a series of steady flow profiles associated with corresponding flow durations. The sediment transport equations are then solved for each time step.

4.1   *Geometry and parameters of the model*

The model is built with sections at a distance of 10 m over the 10.2 km length of the reservoir. The Manning roughness coefficient, n, is equal to 0.05 s/m$^{1/3}$.

In HEC-RAS model, a transport function model needs to be selected by the user. Sediment transport results are strongly dependent on which transport function is selected. Usually when measurements are available, the proper function can be choosen in the model calibra-tion step. In the present study, several functions are tested. Considering the range of assump-tions, hydraulic conditions, and grain sizes, the Toffaleti (Tofaletti, 1968) function is selected. Tofaletti appears to be the most addapted function for modeling suspended load and for the range of grading.

4.2   *Boundary conditions and transport function*

To define the bed load transport capacity as a function of water discharge, different rela-tionships such as modified Meyer-Peter & Müller (MPM) (Meyer-Peter & Müller, 1948), modified MPM (Wong & Parker, 2005) and Smart-Jäggi (1983) are compared. For this purpose a representative section of the river upstream of the reservoir is selected and the solid transport capacity is calculated over a year with a daily time step. The water flow is the mean daily flow. The total bed load sediment volume in one year is then calculated and compared with the 1.7 Mm³ of expected bed load. The modified MPM method by Parker gives a total annual bed load of 1.69 Mm³ and is therefore chosen. In addition, the MPM method is the most well suited for the range of slopes in the Mantaro River and the sediment grading.

4.3 Results

The HEC-RAS model is firstly used to model sediment transport over a mean year. For this purpose the model is run two times, once for the bed load and once for the suspended load. The upstream flow boundary conditions is the flow series with daily values. At downstream, a fixed water elevation equal to the normal water elevation of the dam is considered. The upstream sediment boundary condition is specified as sediment load series for the uppermost section.

As it can be expected, the bed load forms a delta at the entrance of the reservoir. Due to the grain size and the low velocities in the reservoir, the delta cannot move forward downstream and the material accumulates at the upstream limit of the model. However, the suspended material remains on suspension while entering the reservoir and settles down in the its middle.

The simulations start on October 1st and end on September 30th of the next year. As it can be seen in Figure 2 the deposition starts on January and continues over the wet season until end of April. From April to September, during the dry season, the sediment yield is negligible and there is no more deposition in the upper part of the reservoir. However, minor erosions can occur on the upper part of the deposited delta and produce a small amount of material moving downstream.

To better illustrate the same results, a plan view of deposition zones is shown in Figure 3. No sedimentation happens for the first 2'300 m of reservoir upstream. Then the deposited layer height increase to 8.9 meters at a distance of about 6'000 m from the dam. The deposition thickness reduces then to 2 m behind the dam.

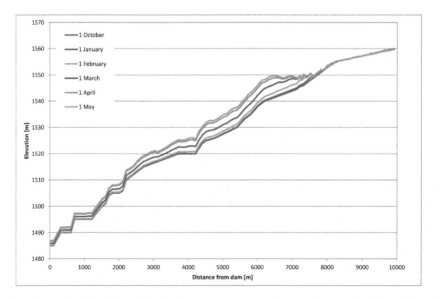

Figure 2.   Suspended load deposition during one year (reservoir normal water elevation 1556 m asl).

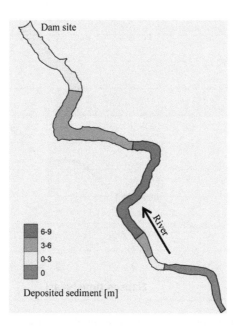

Figure 3. Suspended load deposition zones after one year.

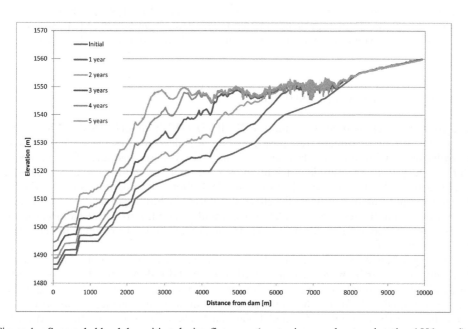

Figure 4. Suspended load deposition during five years (reservoir normal water elevation 1556 m asl).

For the long term deposition, a simulation is carried out over five years. The results are plotted in Figure 4. It is shown that the deposited delta moves on downstream by 2 km each year. A removal measure then seems crucial as the reservoir loses more than half of its capacity in only five years. Figure 5 shows the bed level evolution during 5 years just behind the dam (5 m upstream). As it is shown, during wet seasons high amounts of sediments are deposited behind the dam, whereas for dry seasons there is no deposition.

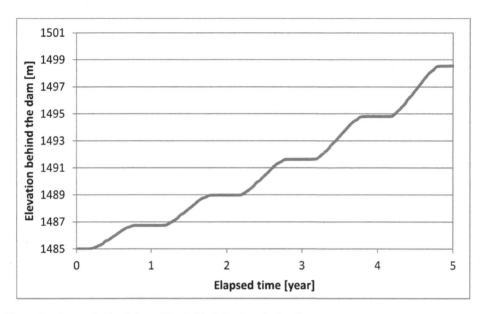

Figure 5.   Suspended load deposition behind the dam during 5 years.

The annual sediment deposition at this section is increased each year comparing to the previous one as the delta approaches the dam. The deposition height at fifth year (about 4 meters) is approximately two times more important than that of the first year (about 2 meters).

5   FULL DRAWDOWN FLUSHING

The entire reservoir is modeled using the 2D BASEMENT model. BASEMENT is well-known for flow and sediment transport modeling. It is developed by the VAW (Laboratory of hydraulics, hydrology and glaciology) of the Swiss Federal Institute of Technology in Zurich. The velocity magnitude distribution over the reservoir and the bed-level shear stress obtained from the model also help to predict deposition zones of transported sediments. The model can be used to evaluate the effect of different hydraulic parameters, such as initial water elevation in the reservoir, or inlet/outlet boundary conditions. Due to long calculation times, it was not used for sediment transport simulations, only clear water simulations are performed instead.

Flushing during annual flood is modeled for different reservoir water surface elevations. The initial reservoir bathymetry is imported from the HEC-RAS model after one year of suspended load deposition (Fig. 2). Sediment transport is not simulated. Only the shear stresses are analyzed during the drawdown. The initial water surface elevation is set to three different levels, 1'556, 1'526 and 1'500 m asl. These elevations are held at these levels throughout the flushing period. The upstream boundary condition is defined as water inflow hydrograph with a peak of 1'400 m³/s which is equal to a flood discharge of 1 year return period.

Figure 6 illustrates the shear stresses at the reservoir bottom for different water surface elevations. For the normal water elevation (1556 m asl) the shear stress at the reservoir bottom is less than 25 N/m² except in the first 3 km. The flow therefore, cannot mobilize the deposited sediments. For water surface elevation of 1526 m asl high shear stresses are obtained for the upstream 7 km of reservoir and more deposited sediment can be mobilized during flushing. Drawing the water surface elevation down to 1500 m asl exposes the whole reservoir bed to

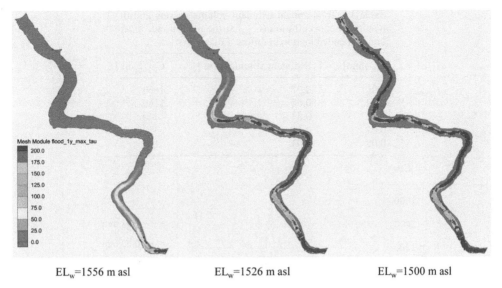

EL$_w$=1556 m asl	EL$_w$=1526 m asl	EL$_w$=1500 m asl

Figure 6. Bed-level shear stress [N/m^2] model during an annual flushing for three different water elevations.

high shear stresses (about 180 N/m^2). The critical shear stress for the suspended load and bead load grains with D$_{50}$ is 1 and 70 N/m^2 respectively. As such, the deposited sediment in the whole reservoir can be mobilized and washed out.

6 SEDIMENT CONCENTRATION IN POWER INTAKES

In order to assess the general distribution of suspended sediment concentrations especially in a main power intake and a small hydropower intake, CE-QUAL-W2 program is used. CE-QUAL-W2 is a two-dimensional, laterally averaged, longitudinal/vertical, hydrodynamic and water quality model. Since the program assumes lateral homogeneity, it is best suited for relatively long and narrow water bodies exhibiting longitudinal and vertical water quality gradients. The model has been applied to rivers, lakes, reservoirs, estuaries, as well as a combination of water bodies (Motamedi, 2012). The model predicts water surface elevation, velocities and temperature being also able to solve transport equations for sediments and inorganic suspended solids. Any combination of constituents can be included in the simulation. To model the suspended load deposition in the reservoir, a settling velocity for the grains is needed. The falling velocity is calculated using the relationships presented by Jimenez & Madsen (2003).

Different long and short term scenarios are modeled to assure a better understanding of suspended material distribution evolution in the reservoir. To get more plausible results, several simulations are performed using a range of grain diameters between 0.01 to 0.05 mm. The total annual sediment volumes passing the main power intakes for suspended load with different diameters are listed in Table 4. The results for concentrations at the main power intake are shown in Figure 7.

The concentrations at the small power intake are slightly lower than those of the main power intake. The results also show that for grains with a nominal diameter more than 0.05 mm the suspended material concentration in the intakes becomes zero. This latter can be explained by very high rate of settling at the upstream part of the reservoir, which causes material deposition at the entrance. As a matter of fact, the suspended material cannot reach the power intakes.

217

Table 4. Total annual sediment volume passing and the maximum concentration (C_{max}) at the main power intake and suspended load with different diameters.

D_{50} [mm]	Sediment volume [Mm³]	C_{max} [g/m³]
0.01	0.97	1660
0.02	0.62	1160
0.03	0.32	740
0.04	0.11	330
0.05	0.03	100

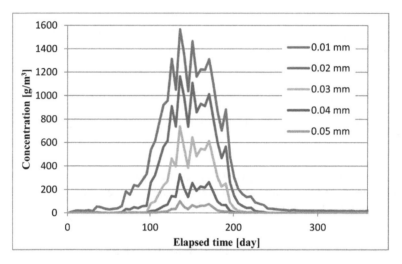

Figure 7. Concentration at power intake for different grain diameters.

7 CONCLUSION

Sediment transport modeling is notoriously difficult. The data used to predict bed changes is fundamentally uncertain and the theory employed is empirical and highly sensitive to a wide range of physical variables. However, with good data, long term trends for planning decisions can be modeled.

When the river reaches the reservoir, due to the increase in water depth in the latter, the flow velocities, turbulences and bed shear stresses are reduced. The bed load part is therefore settled down and forms a delta. The suspended load, however, can be carried by water over a longer distance than the bed load, and the delta that it forms can approach the dam.

The total flushing efficiency of a reservoir is principally guaranteed if a free surface flow can be established during the flushing process.

An important volume of the suspended load is evacuated through the power intakes and then the turbines and does not accumulate in the reservoir. This volume is significantly influenced by the size of the suspended material. For an average size of 0.03 mm, the volume passing through the turbines is estimated to 320'000 m³ annually, which represents 10% of total incoming suspended sediment volume.

REFERENCES

Atkinson, E. 1996. The feasibility of flushing sediment from reservoirs. Report OD137, HR Wallingford, Wallingford, UK.

BASEMENT reference and user manuals, http://www.basement.ethz.ch/.

Bessenasse M., Kettab A., Paquier A., Ramez P. & Galea G. 2003. Simulation numerique de la sédimentation dans les retenues de barrages. Cas de retenue de Zardezas, Algéria, Revue des Sciences de l'eau. 16: 103–122.

CE-QUAL-W2 user manual, http://www.ce.pdx.edu/w2/.

Harb, G., Dorfmann, C., Schneider, J., Haun, S. & Badura, H. 2012. Numerical analysis of sediment transport processes in a reservoir, River Flow conference: 859–865.

HEC-RAS reference and users manuals, http://www.hec.usace.army.mil/software/hec-ras/.

Jimenez J.A. & Madsen O.S. 2003. A simple formula to estimate settling velocity of natural sediments Journal of waterway, port, coastal and ocean engineering: 70–78.

Leite Ribeiro M., De Cesare G., Schleiss A.J. & Kirchen G.F. 2005. Sedimentation management in the Livigno reservoir, International Journal on Hydropower & Dams. 12(6): 84–88.

Meyer-Peter, E; Müller, R. 1948. Formulas for bed-load transport. Proceedings of the 2nd Meeting of the International Association for Hydraulic Structures Research. 39–64.

Möller G., Boes R.M., Theiner D., Fankhauser A., Daneshvari M., De Cesare G. & Schleiss A.J. 2011. Hybrid modeling of sediment management during drawdown of Räterichsboden reservoir, 79th Annual Meeting of ICOLD—Swiss Committee on Dams.

Morris, Gregory L. and Fan, Jiahua. 1998. Reservoir Sedimentation Handbook, McGraw-Hill Book Co.: New York.

Motamedi, K. 2012. Sediment flushing study for Puhulpola Dam. Mahab Ghodss report.

Olsen, N.R.B. 1999. Two-dimensional numerical modelling of flushing processes in water reservoirs, Journal of Hydraulic research. 37(1): 3–16.

Palmieri, A., Shah, F., Annandale, G.W. & Dinar, A. 2003. RESCON, Reservoir Conservation. Volume I.

Shen H.W. 1999. Flushing sediment through reservoirs. Journal of hydraulic research International Association for Hydraulic Research. 37(6): 743–757.

Smart, G.M., Jäggi, M.N.R. 1983. Sediment-transport in steilen Gerinnen. Mitteilungen der Versuchsanstalt für Wasserbau, Hydrologie und Glaziologie der ETH Zürich. 64.

Toffaleti, F.B., 1968. Technical Report No. 5—A Procedure for Computation of Total River Sand Discharge and Detailed Distribution, Bed to Surface. Committee on Channel Stabilization, U.S. Army Corps of Engineers.

White, R. 2001. Evacuation of sediments from reservoirs. Thomas Telford: London.

Wong, M. & Parker, G. 2006. Reanalysis and Correction of Bed-Load Relation of Meyer-Peter and Müller using Their Own Database. Journal of Hydraulic Engineering, (132)11: 1159–1168.

Sediment bypass tunnels to mitigate reservoir sedimentation and restore sediment continuity

R.M. Boes, C. Auel, M. Hagmann & I. Albayrak
Laboratory of Hydraulics, Hydrology and Glaciology (VAW), ETH Zurich, Switzerland

ABSTRACT: Worldwide, a large number of reservoirs impounded by dams are rapidly filling up with sediments. As on a global level the loss of reservoir volume due to sedimentation increases faster than the creation of new storage volume, the sustainability of reservoirs may be questioned if no countermeasures are taken. This paper gives an overview of the amount and the processes of reservoir sedimentation and its impact on dams and reservoirs. Furthermore, sediment bypass tunnels as a countermeasure for small to medium sized reservoirs are discussed with their pros and cons. The issue of hydroabrasion is highlighted, and the main design features to be applied for sediment bypass tunnels are given.

1 INTRODUCTION

By impounding natural watercourses, dams alter the flow regime from flowing water to a body of standing water, which favors reservoir sedimentation. Without adequate countermeasures ongoing sedimentation may lead to various problems such as (1) a decrease of the active volume leading to a loss of energy production or of water available for water supply and irrigation; (2) a decrease of the retention volume in case of flood events; (3) endangerment of operating safety due to blockage of the outlet structures; and (4) increased abrasion of steel hydraulics works and mechanical equipment due to increasing specific suspended load concentrations. Besides these operational problems a lack of sediments in the downstream river stretch may result in river bed incision. Particularly, with more severe legislation such as the revised Swiss water protection law that has come into force in 2011, the exigencies regarding ecology have increased. One of the goals is to restore the longitudinal continuity of sediments wherever possible at reasonable expense. For many smaller reservoirs, particularly in mountainous conditions, Sediment Bypass Tunnels (SBTs) may counter these negative effects by connecting the upstream and downstream reaches of dams and reestablishing sediment continuity, as proven by a number of cases worldwide, particularly in Japan and Switzerland (Auel & Boes 2011, Fukuda et al. 2012). However, due to high flow velocities and large bed load transport rates, hydroabrasion is a frequent phenomenon present at SBT. Due to the fact that abrasion requires continuous maintenance and causes high annual costs, adequate countermeasures such as using High-Performance Concrete (HPC) and/or optimization of hydraulic conditions for invert protection should be already taken into account at the design phase (Hagmann et al. 2012).

2 RESERVOIR SEDIMENTATION AND COUNTERMEASURES

In analogy to natural lakes, artificial reservoirs impounded by dams fill up with sediments over time. Depending on local site conditions such as size, topography, landform, hydrology and geology of the catchment basin, as well as size and shape of the reservoir this process may last from a few years to several centuries. On a worldwide scale, typical sedimentation rates per country vary between a few tenths up to more than three percent (Fig. 1),

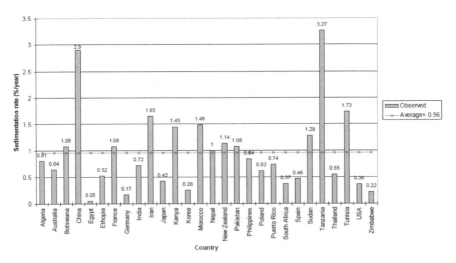

Figure 1. Observed sedimentation rates for various countries worldwide (after ICOLD 2009).

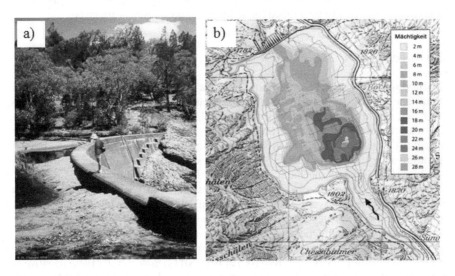

Figure 2. (a) Fully-silted *Koorawatha* reservoir in Australia (Chanson 1998), (b) aggradation depths in *Räterichsboden* reservoir from 1950 to 2001 (Bühler & Anselmetti 2003).

putting the sustainability of reservoirs into question if no adequate countermeasures are taken. The Koorawatha reservoir in Australia, for instance, has quickly experienced considerable sedimentation after commissioning in 1911, so that it lost its main purpose for railway water supply (Fig. 2a). Although such a high rate of sedimentation hardly occurs for Central European conditions, some alpine reservoirs also show a significant aggradation process. The aggradation depths in the Räterichsboden reservoir in the Swiss Alps amounted to 28 m after 50 years of operation (Fig. 2b).

From a hydraulics and sedimentology point of view the deposition process of bed load or suspended load in a reservoir is described by the relationship between discharge, flow velocity or bed shear stress and particle properties e.g. size, density and settling velocity. The aggradation pattern in a reservoir therefore depends on the kind and amount of incoming sediments as well as the geometry and operation mode of the reservoir. Typically, due to decreasing flow velocities and thus turbulence intensities aggraded sediments become finer from the

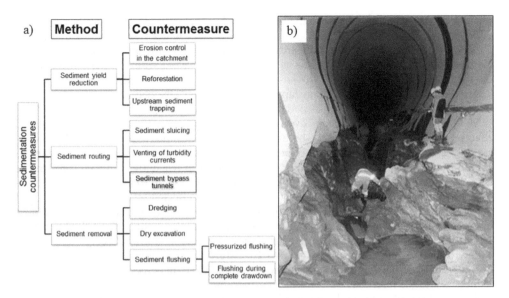

Figure 3. (a) Reservoir sedimentation countermeasures (adapted from Sumi et al. 2004), (b) invert abrasion at *Palagnedra* SBT, Switzerland (photograph by C. Auel).

upper reach of the reservoir towards the dam. A common approach distinguishes between delta formation at the upper reach of a reservoir caused by coarse sediments (bed load) and the aggradation of fines in the deeper water zone further downstream which is often highly affected by density currents in the case of rather narrow and elongated reservoirs of steep bottom slope (Schleiss et al. 2010, Boes 2011).

Reservoir sedimentation causes a number of negative impacts on dams. Firstly, when reaching the dam it may endanger the functionality of both intake structures and bottom outlets. Blockage of the latter must be avoided, as these constitute an important safety element of dams. Secondly, the effective net volume available for the purpose of the reservoir, e.g. power production or flood protection, is reduced over time due to proceeding sedimentation. Thus aggradation of fines results in an immediate negative impact, whereas accumulation of coarse material has a long-term negative impact on reservoirs.

To keep or restore the original reservoir volume the necessary measures are (I) prevention of sediment input, (II) routing of incoming sediments and (III) removing aggraded sediments *a posteriori* (Fig. 3a). Whereas the former have a preventive character, i.e. they impede sediments from being transported into a reservoir; the two latter methods are retroactive, as they deal with sediments that have already been transported into the lake. Sediment bypass tunnels belong to the routing method, as they convey sediments around a dam into the tailwater. SBTs are mainly operated during flood events and connect the upper and lower river reaches and reestablish the pre-dam conditions in terms of sediment transport (sediment continuity). In general, such measures should be taken as early as possible to maximize their efficiencies, i.e. in the planning and design phases of dams and reservoirs. Unfortunately, despite knowledge on the reservoir sedimentation process countermeasures have often been postponed or not adequately been considered in the past, restricting the choice of efficient measures at a later stage.

3   SEDIMENT BYPASS TUNNELS

According to Auel & Boes (2011) a SBT consists of a guiding structure installed in the reservoir, an intake structure with a gate, mostly a short and steeply sloped acceleration section, a mild sloped bypass tunnel section, and an outlet structure. Depending on the location of the intake structure, i.e. whether at the head or within the reservoir, there are basically two different types of SBTs. For type A, the inflow takes place under free surface conditions at the delta, while for type B it is usually pressurized located below the pivot point of the aggradation body. The tunnel invert has to be steep enough to avoid sediment deposition and at the same time it should be as mild as possible to limit the flow velocities in order to prevent invert abrasion. For existing SBTs in Japan and Switzerland, the bed slope varies between 1 and 4% (Auel & Boes 2011).

SBTs feature several advantages over other countermeasures. Firstly, they have positive effects regarding ecological aspects, because sediment conveyance may significantly decelerate or even stop river bed erosion and increase the morphological variability downstream of a dam. Mainly sediments provided from the upstream river reach are conveyed through the SBT since remobilization of accumulated sediments in the reservoir hardly occurs. The sediment concentration in the tailwater of the dam is thus not affected by the reservoir itself and of natural character. Secondly, SBTs have been proven as an effective countermeasure against reservoir sedimentation amongst others. For instance, the type A SBT of the Asahi Dam in Japan has greatly reduced the severe aggradation in terms of accumulated sedimentation volume after commissioning in 1998 (Fig. 4). Even during an exceptionally large flood caused by a typhoon in 2011 the routing of sediments around the dam helped to limit the inflow of sediments into the reservoir.

Whereas typically, the bed load deposition may be completely solved with SBT, the deposition of fines depends on the design discharge of the tunnel. The higher the recurrence interval of the SBT operation, the higher the share of the incoming suspended load that may be conveyed through the tunnel and the smaller the amount of fines entering the reservoir. The main drawback of SBT is related to economic considerations. The implementation of SBT is not only costly from an investment perspective, but also requires regular maintenance. Due to high flow velocities with peaks in the range of 12 to 20 m/s (Auel & Boes 2011) and high sediment transport rates, invert abrasion is generally a severe problem, requiring costly repair works and maintenance (see Fig. 3b). For this reason SBTs should be considered as a convenient measure for small to medium-sized reservoirs with capacity-to-inflow ratios, i.e. the ratio between the annual inflow and the total reservoir volumes, of about 0.003 to 0.2 for the Swiss and Japanese SBTs (Sumi & Kantoush 2011). For such reservoirs bed load aggradation and delta formation are more critical problems than the problem of sedimentation of fines since a large amount of incoming fines stay in suspension due to the relatively short residence time and are discharged via the outlet works. Moreover, major tunneling costs favor their use at smaller reservoirs due to short SBT lengths.

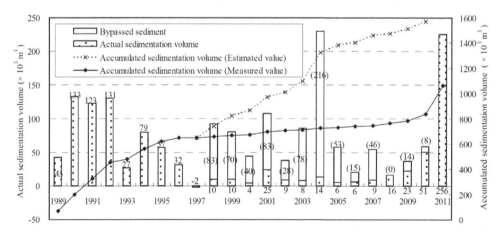

Figure 4. Development of reservoir sedimentation volume at *Asahi* reservoir, Japan, prior to and after commissioning of an SBT in 1998 (Fukuroi 2012).

As stated above, abrasion is a serious concern in most SBTs. The extent of damages along the wetted perimeter, i.e. mainly on the invert and the lower parts of the tunnel walls, typically increases with increasing unit sediment load, particle size and distribution and flow velocities i.e. shear stress as well as quartz content in the mineralogical composition of the sediments.

Hydro-abrasive damage on an invert of a hydraulic structure occurs when the flow induced bed shear stress exceeds a critical value and hence numerous particles start impacting. Depending on the flow conditions particles are transported in sliding, rolling or saltation modes and cause grinding, rolling or saltating impact stress on the bed and thus wear on the bed (Fig. 5). According to Sklar & Dietrich (2001, 2004) the governing process causing abrasion is saltation, whereas sliding and rolling do not cause significant wear. Therefore, for an optimum SBT design, hydraulic conditions, particle size and distribution and hence particle transport modes must be determined. The rolling, saltation and suspension probabilities of different particle sizes are also important since the transport mode directly affects the particle impact energy on the bed. They are determined using the relationship between probabilities of transport modes and Shields number (Hu & Hui 1996, Ancey et al. 2002, Auel et al. 2014a). For a given particle size and flow induced shear stress, particle transport modes and their probability as well as expected particle abrasion mechanisms can be obtained to optimize the hydraulic condition and to choose the invert material type for a SBT.

Auel et al. (2014b) show that mean particle and impact velocities for rolling and saltating particles linearly increase with flow velocity independent of particle size. Large particles possess a large particle mass, thus higher impact energy is transferred by these particles on the surface (Auel et al. 2014a and 2014b). Consequently, a combination of large particle size and high flow velocities results in high mean invert abrasion, which is clearly confirmed in Figure 6. Depending on the invert material and transported sediment properties, the abrasion depth varies and in general typical mean abrasion depths range from microns to some millimeter per operating hour for some Swiss and Japanese cases (Jacobs et al. 2001, Kataoka 2003, Sumi et al. 2004, and Fukuroi 2012).

The results from *Runcahez* SBT indicate that mean abrasion depth per hour increase with decreasing compressive and bending tensile strengths of the invert concrete (Figs. 6a and 7). Moreover, despite slightly larger particle size and higher design velocity compared to *Runcahez* SBT, the mean abrasion depth per hour on *Pfaffensprung* SBT is smaller than in *Runcahez*. This result reveals a strong effect of invert material properties on abrasion depth, e.g. the higher the compressive and bending tensile strengths, the less the abrasion depth (Fig. 7). Comparison of *Runcahez* SBT with *Asahi* SBT shows the effect of design velocity on abrasion depth (Fig. 6b). Although the concrete compressive strength is smaller for *Asahi* than for *Runcahez* (Fig. 7a), the high mean abrasion depth per hour in *Asahi* SBT in comparison to Runcahez SBT is mostly attributed to the high design velocity despite much smaller particle size, i.e. $d_m = 50$ mm for *Asahi* vs. 225 mm for *Runcahez* (Fig. 6a). Note that the bed-load rates are not known for both SBTs.

One of the important findings from the study in *Pfaffensprung* is the high hydroabrasion resistance of granite compared to the implemented concrete invert, as the mean abrasion per hour on the granite plates is five times less than on the concrete invert (Fig. 7a). This may

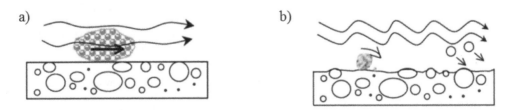

Figure 5. Abrasion processes of hydraulic systems' surfaces: (a) grinding, (b) combination of grinding and impingement (Jacobs et al. 2001).

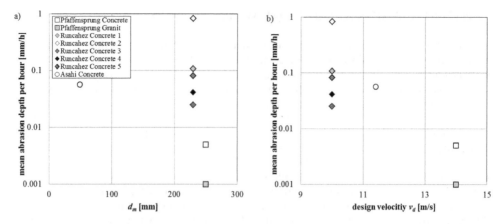

Figure 6.   Observed mean abrasion depths per operating hour as functions of (a) particle diameter d_m and (b) design flow velocity v_d.

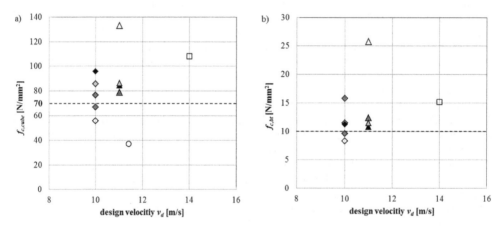

Figure 7.   (a) Compressive strength determined at cubes and (b) bending tensile strength of invert concrete applied in various SBTs (data from studies mentioned above); symbols are the same as in Figure 6.

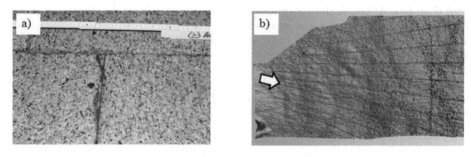

Figure 8.   Abrasion patterns at *Pfaffensprung* SBT: (a) grooves forming along joints of granite plates, (b) undulating invert at HPC test field (steel-fibre HPC with $f_c > 70$ N/mm²).

suggest that granite is a better choice as an invert material over HPC. However, the cost of granite should be carefully considered at SBT design phase.

The results of the study carried out in *Pfaffensprung* SBT in 2013 show that the damages typically take place in the form of grooves along the joints of basalt and granite plates (Fig. 8a), while a wavy pattern of abrasion occurs on the HPC (Fig. 8b). In order to further reduce the abrasion on the granite, this result suggests that such plates should not be implemented in parallel to the main flow direction.

Whereas natural stone material, e.g. granite or cast basalt plates, usually have a high abrasion resistance against pure particle grinding action, their brittleness favors fracturing by impinging particles in case of saltating sediments. In the latter case, either steel or cementitious material such as HPC generally show a better resistance. As steel linings are often too costly for abrasion protection of large areas such as in SBTs, HPC becomes an interesting and economical alternative.

Based on a long-term field study performed at *Runcahez* SBT in Switzerland between 1995 and 2000 (Jacobs et al. 2001), the decisive material characteristics of hardened HPC are the compressive strength ($f_c > 70$ N/mm^2 at 28 days, Fig. 7a), the bending tensile strength ($f_{c,bt} > 10$ N/mm^2 at 28 days, Fig. 7b), and the fracture energy (>200 J/m^2 at 28 days).

5 DESIGN OF SBT

To reduce the negative effects of hydroabrasion at SBTs, (i) an optimum hydraulic design to limit the strong particle impact forces and (ii) a selection of sustainable and optimum abrasion-resistant invert lining material are recommended measures. Aspect (i) demands for the following, among others:

- A tunnel cross section with plane invert geometry should be chosen, i.e. archway and horseshoe profiles with horizontal bed rather than circular ones, to avoid stress concentrations.
- Whenever possible, bends in plan view should be avoided to reduce shock waves and secondary currents, which cause locally high specific sediment transport rates. For instance, in Solis SBT the sediment transport at the tunnel outlet is concentrated on the orographic right side as a result of a bend further upstream (Fig. 9a).
- Keep the bed slope as mild as possible without endangering sediment aggradation.

As to aspect (ii) the following should be accounted for, among others:

- If most of the particles are transported in rolling or sliding motion with only minor saltation, abrasion processes are expected to be mainly grinding and only weakly impinging. Hence using natural stones such as granite and cast basalt as invert lining material is a good solution. As the joints should not be parallel to the flow, the use of hexagonal plates is recommended (Fig. 9b). The plates should be embedded into a special mortar.
- If saltation is expected to be the main particle transport mode and/or the sediment is rather coarse with high flow velocity, HPC with compressive strength above 70 N/mm^2 (i.e. C70/85 and higher) and a bending tensile strength above 10 N/mm^2 is preferable. Concrete curing is critical and should be carefully performed. For more details, see Jacobs et al. (2001).

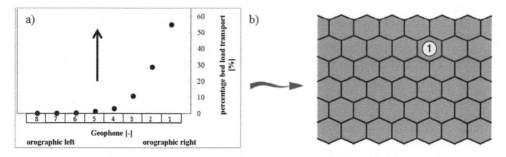

Figure 9.   (a) Measured sediment transport using 8 geophones for the 2013 flood event at *Solis* SBT; (b) schematic plan view of a thin pavement of hexagonal natural stone plates (Jacobs et al. 2001).

6 CONCLUSIONS

Reservoir sedimentation is a serious worldwide problem threatening the sustainability of reservoirs and negatively impacting dam safety. SBTs are effective countermeasures for small to medium-sized reservoirs provided that hydroabrasion is accounted for from the very beginning of the planning stage by optimizing the hydraulic design and by applying adequate invert material. Depending on the transport mode of the sediment which is determined by the hydraulic characteristics of the tunnel flow, cast basalt and granite plates or HPC and steel linings are recommended.

REFERENCES

Ancey, C., Bigillon, F., Frey, P., Lanier, J. & Ducret, R. 2002. Saltating motion of a bead in a rapid water stream. *Physical Review E* 66(3).

Auel, C., Albayrak, I. & Boes, R.M. 2014a. Particle motion in supercritical open channel flows. *Journal of Hydraulic Engineering*: submitted.

Auel, C., Albayrak, I. & Boes, R.M. 2014b. Bedload particle velocity in supercritical open channel flows. *Proc. Intl. River Flow Conference,* EPF Lausanne, Switzerland: submitted.

Auel, C. & Boes, R.M. 2011: Sediment bypass tunnel design—review and outlook. *Proc. ICOLD Symposium "Dams under changing challenges"* (A.J. Schleiss & R.M. Boes, eds.), 79th Annual Meeting, Lucerne. Taylor & Francis, London: 403–412.

Boes, R. 2011. Nachhaltigkeit von Talsperren angesichts der Stauraumverlandung ('Sustainability of dams in view of reservoir sedimentation'). *Mitteilung 164,* Lehrstuhl und Institut für Wasserbau und Wasserwirtschaft (H. Schüttrumpf, ed.), RWTH Aachen, Germany: 161–174 [in German].

Bühler, R. & Anselmetti, F. 2003. Ablagerungen in den Grimsel-Stauseen. Teil B: Räterichsbodensee ('Deposition in the Grimsel reservoirs. Part B: Räterichsboden reservoir'). *Report* Limnogeology lab of the Geological Institute, ETH Zurich, Switzerland [in German].

Chanson, H. 1998. Extreme reservoir sedimentation in Australia: a Review. *Intl Jl. of Sediment Research* 13(3): 55–63 (ISSN 1001-6279).

Fukuda, T., Yamashita, K., Osada, K. & Fukuoka, S. 2012. Study on Flushing Mechanism of Dam Reservoir Sedimentation and Recovery of Riffle-Pool in Downstream Reach by a Flushing Bypass Tunnel. *Proc. Intl. Symposium on Dams for a changing world,* Kyoto, Japan.

Fukuroi, H. 2012. Damage from Typhoon Talas to Civil Engineering Structures for Hydropower and the Effect of the Sediment Bypass System at Asahi Dam. *Proc. Int. Symposium on Dams for a changing World—Need for Knowledge Transfer across the Generations and the World.* Kyoto, Japan.

Hagmann, M., Albayrak, I. & Boes, R.M. 2012. Reduktion der Hydroabrasion bei Sedimentumleitstollen—In-situ-Versuche zur Optimierung der Abrasionsresistenz ('Reduction of hydroabrasion in sediment bypass tunnels—in-situ experiments to optimize abrasion resistance'). *Proc. Wasserbausymposium,* TU Graz (G. Zenz, ed.), A12: 91–98 [in German].

Hu, C. & Hui, Y. 1996. Bed-load transport. I: Mechanical characteristics. *Journal of Hydraulic Engineering* 122(5), 245–254.

ICOLD 2009. Sedimentation and Sustainable Use of Reservoirs and River Systems. *Draft Bulletin 147,* Paris, France.

Jacobs, F., Winkler, K., Hunkeler, F. & Volkart, P. 2001. Betonabrasion im Wasserbau ('Concrete abrasion at hydraulic structures'). *VAW-Mitteilung 168* (H.-E. Minor, ed.), VAW, ETH Zurich, Switzerland [in German].

Kataoka, K. 2003. Sedimentation Management at Asahi Dam. *Proc. Third World Water Forum,* Siga, Japan, 197–207.

Schleiss, A., De Cesare, G. & Jenzer Althaus, J. 2010. Verlandung der Stauseen gefährdet die nachhaltige Nutzung der Wasserkraft ('Reservoir sedimentation threatens the sustainable use of hydropower'), *Wasser, Energie, Luft,* 102(1): 31–40 [in German].

Sklar, L.S. & Dietrich, W.E. 2001. Sediment and rock strength controls on river incision into bedrock. *Geology* 29(12), 1087–1090.

Sklar, L.S. & Dietrich, W.E. 2004. A mechanistic model for river incision into bedrock by saltating bed load. *Water Resources Research* 40(W06301), 21p.

Sumi, T. & Kantoush, S.A. 2011. Comprehensive Sediment Management Strategies in Japan: Sediment bypass tunnels. *Proc. 34th IAHR World Congress.* Brisbane, Australia, 1803–1810.

Sumi, T., Okano, M. & Takata, Y. 2004. Reservoir sedimentation management with bypass tunnels in Japan. *Proc. 9th International Symposium on River Sedimentation.* Yichang, China, 1036–1043.

Reservoir Sedimentation – Schleiss et al. (Eds)
© 2014 Taylor & Francis Group, London, ISBN 978-1-138-02675-9

Modeling long term alternatives for sustainable sediment management using operational sediment transport rules

S.A. Gibson
Hydrologic Engineering Center, U.S. Army Corps of Engineers, Davis, CA, USA

P. Boyd
U.S. Army Corps of Engineers—Omaha District, Omaha, NE, USA

ABSTRACT: Specifying reservoir operations as an *a priori* boundary condition complicates long term, predictive, sediment modeling because reservoir hydrology, hydraulics, and sedimentation interact in morphological simulations and each has feedbacks on structure operations. Determining future operations iteratively can be difficult and time consuming, particularly for multi-decadal or stochastic models. Automatically updating structure operations mid-simulation in response to intermediate model results improves the efficiency and accuracy of predictive reservoir sediment modeling.

Version 5.0 of the Hydrologic Modeling Center's River Analysis System (HEC-RAS) couples automated, responsive, operation options with the sediment model, facilitating long term, predictive flushing and routing simulations. HEC-RAS can activate or delay flushing events, routing operation, or dredging in response to mid-simulation results and gradients. This paper demonstrates the advantage of new features in HEC-RAS 5.0 for modeling sustainable reservoir sediment management options and documents recent US Army Corps of Engineers applications on the Missouri River and Tuttle Creek.

1 INTRODUCTION

One dimensional (1D) sediment transport models can be effective tools for flushing and flushing analysis. Multi-dimensional and physical models offer higher fidelity results but classical model complexity tradeoffs (Gibson, 2013a) like run time, cost, and equifinality (Bevan, 1993) make 1D models attractive options for preliminary assessment, stochastic uncertainty analysis, and for screening a broad range of alternatives, particularly on the multi-decadal time scales relevant to reservoir life cycle analysis. Earlier versions of the US Army Corps of Engineers' (USACE) 1D sediment transport model HEC-RAS have been applied successfully to sustainable sediment management studies (Davis *et al.*, 2014). However, limitations encountered in early attempts to apply HEC-RAS to long term sediment management studies (e.g. flushing, routing, and dredging) drove development of a suite of capabilities targeted to improve the efficiency and fidelity these studies. This paper introduces two recent USACE sustainable sediment management studies and highlights the utility of new capabilities developed for these studies, including implementation of Operational Rules.

2 USACE SUSTAINABLE SEDIMENT MANAGEMENT STUDIES

Sediment flushing and routing are two dominant operational alternatives for sustainable sediment management studies. According to Morris and Fan's (1998) classic taxonomy of sustainable sediment management alternatives, "flushing" alternatives draw down the reservoir to run of river conditions to scour historic reservoir deposits while "routing" operations pass sediment laden flows through the reservoir before they have an opportunity to deposit.

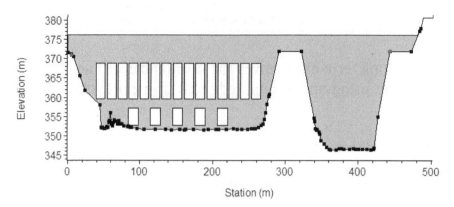

Figure 1. Alternative Gavins Point Dam Spillway geometry modeled with low level outlets.

A sediment *flushing* study on the Lewis and Clark Lake and a sediment *routing* analysis for Tuttle Reservoir demonstrated the limitations of classic 1D approaches for flushing and routing studies and provided the opportunity to advance HEC-RAS model capabilities.

The Lewis and Clark Lake is reservoir behind Gavins Point Dam near Yankton, South Dakota. It is the downstream most pool of a 1500 km, six reservoir Missouri River cascade in the Omaha District of the USACE, which spans the upstream two thirds of the Missouri River and includes several of the largest reservoirs in the world. The upstream reservoirs intercept all appreciable main stem sediment load before it reaches Lewis and Clark Lake. Most sediment delivered to the reservoir comes from the Niobrara River, a tributary 50 km upstream of the dam that drains the Sandhills region of Nebraska and delivers about 5 million metric tons per year, mostly fine sand. As of 2012, 25% of the total reservoir storage has been lost to sedimentation since its closure in 1955. There are three listed species in the downstream reach that are considered impacted by decreased sand loads (e.g. loss of emergent sandbar habitat and shallow water habitat).The dominance of sand in the reservoir and the distance between the delta and the reservoir preclude routing alternatives. But HEC-RAS was used to investigate flushing options. Ten alternatives including drawdown hydrographs of different duration and magnitude were modeled independently and in conjunction with structural gate modifications (e.g. low level outlets added in Fig. 1), revetments and dredging schedules.

Tuttle Reservoir, in central Kansas (managed by the Kansas City District of the USACE) is a very different system. Fine sediment (<63 μm) has displaced almost 40% of the reservoir volume since closure. In 2013 a multi-agency panel decided that routing alternatives (Morris and Fan, 1998) were the most promising and proposed to model scenarios that pass the sediment laden flows on the rising limb of the hydrograph before they have the opportunity to deposit. An unsteady HEC-RAS sediment model was identified to analyze these reservoir operations. The distinct processes and objectives of this study provided an opportunity to further develop and test reservoir modeling capabilities.

3 NEW MODEL CAPABILITIES

Several new features were added to HEC-RAS 5.0 to supporting sustainable reservoir sediment management studies. These features group into two main categories: 1) improvements to hydrodynamics and structure operation flexibility—particularly the capability to model mid-simulation anthropogenic feedback with operational rules, and 2) improvements to the reservoir erosion algorithms.

3.1 *Unsteady transport and operational rules*

Earlier versions of HEC-RAS modeled sediment transport hydrodynamics with a 'quasi-unsteady' flow model (Gibson *et al.*, 2006), approximating hydrographs by attaching temporal information to a series of steady flows. This is a common hydrodynamic simplification in sediment models, regardless of dimensionality, and usually offers an attractive tradeoff. Avoiding the hyperbolic unsteady flow equations improves model stability while the assumption generally introduces very modest errors due to the time scale difference between hydrodynamic and sediment system response. However, the qusi-unsteady model performs particularly poorly for systems with substantial hydrologic storage. This limitation makes the quasi-unsteady assumption problematic for many reservoirs simulations, and particularly inappropriate for rapid drawdown and refilling associated flushing and routing studies.

Therefore, the sediment features were coupled with the unsteady flow solver and interface. HEC-RAS uses a Preissman box scheme to solve the Saint Venant equation implicitly using a skyline matrix solver or the PARDISO library:

$$\frac{\partial A}{\partial t} + \frac{\partial(\Phi Q)}{\partial x_c} + \frac{\partial[(1-\Phi)Q]}{\partial x_f} = 0 \tag{1}$$

$$\frac{\partial Q}{\partial t} + \frac{\partial(\Phi^2 Q^2 / A_c)}{\partial x_c} + \frac{\partial[(1-\Phi)^2 Q^2 / A_f]}{\partial x_f} + gA_c\left[\frac{\partial Z}{\partial x_c} + S_{fc}\right] + gA_f\left[\frac{\partial Z}{\partial x_f} + S_{ff}\right] = 0 \tag{2}$$

where Q = flow, A = area, S_f = friction slope, z = water surface elevation, $\Phi = (K_c/(K_c-K_f))$, K_c = channel conveyance K_f = floodplin conveyance and subscripts c and f refer to the channel and floodplain respectively. Sediment is routed and bed change computed after each unsteady time step using the Exner equation:

$$(1-\lambda_p)B\frac{\partial \eta}{\partial t} = \frac{\partial(Q_s)}{\partial x} \tag{3}$$

where B = channel width, η = channel elevation, λ_p = porosity, and Q_s = transported sediment load. Transported sediment is computed with one of eight transport capacity methods and bed dynamics are simulated with bed mixing algorithms.

However the Exner equation applies no temporal control to transport. Computing realistic residence times are essential to simulating reservoir deposition accurately and the simple continuity solution of the Exner equation in previous versions of the model did not have any mechanistic temporal routing component. Therefore, *ad hoc* temporal limiters (Gibson, 2011) was implemented to limit sediment transport to water velocity and a first order solution of the Advection-Dispersion (AD) equation is included in developmental versions of the code.

Solving sediment transport after each time step of implicit, unsteady, Saint-Venant solution is not intrinsically innovative. Many research and production sediment codes have this capability. However, coupling the sediment model to the unsteady capabilities in HEC-RAS is advantageous for long term reservoir sedimentation simulations. This connection leverages a suite of powerful and flexible generalized tools native to the mature unsteady flow modeling environment for reservoir sediment studies. These includes lateral structures, extensive visualization tool, and the Internal Flow Stage Boundary condition ("IB Stage/Flow") which computes the necessary flow through a structure to reproduce a historic reservoir stage series (or vice versa). However, the "Operational Rules" capability has proven particularly useful for routing and flushing studies.

Operational Rules enable gate operations giving modelers flexibility to automate their reservoir operations based on operational objectives and constraints as a function of mid-simulation results. User defined equations, simple logical code, or complex operational algorithms specify mid-simulation changes to structure flows or gate operations in response to flows, stages, or structure operations elsewhere in the model (e.g. operating one structure in response to operations on an upstream structure). For example an 8.5 day, 1700 cms, run of river flushing event was defined for Lewis and Clark Lake with the relatively simple set of rules in Figure 2. Similar operational rules were applied to alternatives that included only existing gates or that optimized flow between the existing gates and proposed low flow outlets (Fig. 1). These parsimonious rules generated the relatively complex flushing hydrograph in Figure 3 and the sediment volume change at each cross section in Figure 4.

Results (Fig. 4) demonstrate that low flow outlets did not increase scour appreciably in the reservoir delta (stations 250–253 km) but they did preclude re-deposition in the pool. However, regardless of gate configuration, most of the material removed by this event is fine (i.e. silt or clay).

While the rules capability makes constructing flushing events without violating social, physical or numerical constraints faster and easier, the tool is more powerful for long term simulations where the complexity of the hydrograph and the nature of the feedbacks make determining a hydrograph *a priori* almost impossible. A slightly more complex set of rules (than Fig. 2) introduced annual flushing events to a period of record simulation when calendar, stage and flow constraints were met. While a tedious iterative approach could be used

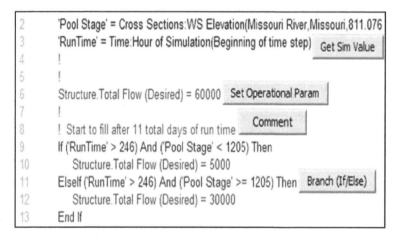

Figure 2. Sample code in HEC-RAS Operational Rules editor which generated the flushing hydrograph in Figure 3 for Lewis and Clark Reservoir. The buttons associate with each basic 'type' of command is also included.

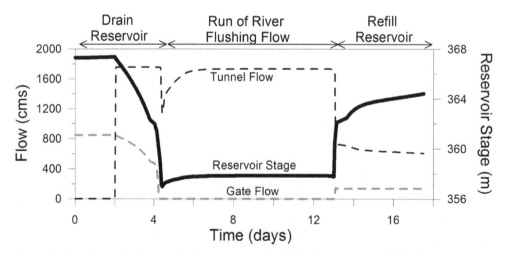

Figure 3. Lewis and Clark Lake flushing hydrograph with the low outlets added in Figure 1 and the operational rules defined in Figure 2.

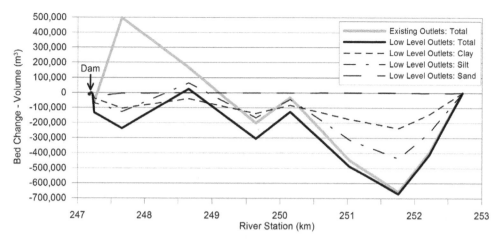

Figure 4. Bed change volume at each cross section for an 8.5 day, 1,700 cms run of river flushing event with and without the low level outlets. The low outlet results are subdivided by grain class. Dam is just upstream of station 247.

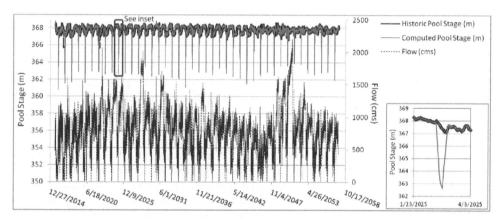

Figure 5. Fifty-Seven year period of record simulation starting with the 2011 geometry, which uses operational rules to maintain the historic water surface elevation coupled with the historic flows, but introduces an 850 cms flushing event every spring based on calendar and physical constraints.

to eventually develop the hydrograph in Figure 3, it would be nearly impossible to develop the long term hydrograph Figure 5, particularly when the sediment feedbacks on reservoir volume and stage are considered. This simulation estimated an annual, one week, 850 cms flush would reduce deposition by approximately 600 million m³. The profile of depositional difference is plotted in Figure 6.

Operational rules will also improve the efficiency and accuracy of the sediment flushing model of Tuttle reservoir. Rules set to monitor the absolute flow and rate of change at the upstream boundary condition also facilitate long term flushing simulations that draw down reservoirs to pass the rising limb of the hydrograph (and the most sediment laden flows), and then evaluate alternatives for timing these operations. Again, an iterative approach could eventually yield single event results of this sort, but the ability to specify mid-simulation anthropogenic responses to physical processes and thresholds in multi-decadal simulations is a powerful tool for these studies.

In addition to the hydrologic and hydraulic flexibility the operational rules offer flushing and routing analyses, HEC is actively adding sediment parameters to the operational rules.

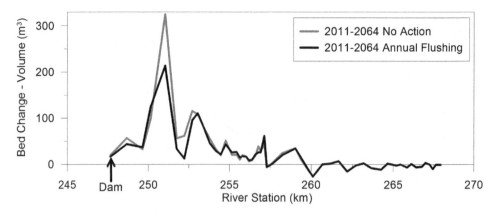

Figure 6.   Bed change volume at each cross section for a fifty-seven year, period of record reservoir sediment transport simulation with and without annual, one week, 850 cms flushing events.

This provides modelers the opportunity to automate reservoir operations based on sediment processed. For example:

- If the concentration of water released downstream of the structure exceeds a TMDL or limit specified by a resource agency close the gates or limit the flow through the structure.
- If the sediment elevation at a location (e.g. infrastructure location) increases to a specified elevation, initiate a flushing event when the hydrologic, sediment, and calendar constraints next allow it.
- If the inflowing sediment concentration exceeds a threshold in a long term period of record simulation, open the gates to route the high concentration water before it settles.

3.2   *Reservoir erosion model*

Distributing bed change across a cross section is a challenge for 1D sediment models. These models generally convert sediment surplus or deficit into cross section change with some variation of the "veneer method," adjusting all wetted nodes (within user specified lateral limits) an equal vertical distance to reflect the computed mass change. While the veneer method is often reasonable and there are several useful variations on this theme, its limitations posed a second major challenge to using a 1D sediment model particularly for flushing analysis. Reservoir deltas and deposits usually span the lateral extent of the flooded valley, but flushing alternatives tend to erode narrow channels (Morris and Fan, 1998, Morris *et al.*, 2007) though an accelerated version of the basic principles of channel evolution (Schumm *et al.*, 1984). Eroding a narrow channel has computational feedbacks on morphology, increasing water depth, which increases shear and total scour at the cross section. Therefore, the veneer method over-predicts the width which can under predict the rate of reservoir scour in response to a run-of-river drawdown event.

The 1D veneer limitation has analog in dam removal modeling. Mature algorithms have been developed for applying a simplified channel evolution model to improve 1D modeling of reservoir sediment in these applications (Cantelli *et al.*, 2004, Gaiman and Huang, 2006, Cui *et al.*, 2006). Because a run of river flushing event is, conceptually, a temporary dam removal, these methods can be leveraged for sediment management analysis. Therefore a simplified channel evolution model was added to HEC-RAS. Adding a simplified channel evolution model to a sediment code is relatively straight forward compared to parameterizing it. The model is parsimonious, with only two parameters: maximum channel bottom width and channel side slope. However, both parameters are notoriously difficult to estimate. The Atkinson (1996) equation that regresses channel widths observed during reservoir drawdowns against flow was added as a calculator, but both parameters require sound engineering intuition and sensitivity analysis. If a more detailed analysis of channel development in

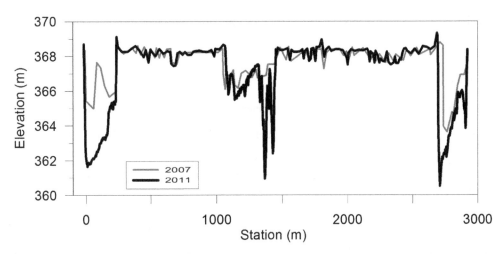

Figure 7. Between 2007 and 2011 the three main reservoir delta channels scoured while the vegetatively stabilized islands between them did not.

reservoir sediment is required, the USDA Agricultural Research Service (ARS)-Bank Stability and Toe Erosion Model (BSTEM) (Simon *et al.*, 2000, Langendoen and Simon, 2008) has been added to HEC-RAS 5.0 (Gibson, 2013, Boyd and Gibson, 2013). The BSTEM tool in HEC-RAS simulates channel development mechanistically and could help parameterize the simple evolution model (Gibson, 2013, Boyd and Gibson, 2013).

Additionally, drawdown events in large reservoirs often scour multiple channels in delta sediment, which can pose challenges to applying either the veneer method or a simple channel evolution model. This process was encountered in Lewis and Clark Lake, where water and sediment were transported primarily by two peripheral channels in the upper delta, which branched to include a third central distributary as it approached the foreset slope. During the historic 2011 flood on the Missouri River, high flows were not coupled with large sediment loads (because the local tributary that delivers most of the sediment did not experience a coincident flood). During these high flow-low load conditions the delta scoured and repeated surveys provide prototype evidence of how the delta may respond to a scouring event. Most of the incision was confined to these channels while the 'islands' between them stayed static (Fig. 7). Therefore, neither the veneer method nor a simple channel evolution model could simulate this process. Multiple erodible limits, which divided the cross section into multiple zones of erodible and non-erodible regions were added to simulate this process, and improved calibration against the 2011 scour event.

4   CONCLUSION

HEC-RAS 5.0 includes new features that facilitate sustainable reservoir sediment management analysis. The suite of unsteady tools in the model can now be applied to sediment studies. The operational rules, which define adaptive reservoir operation based on computed mid-simulation results are particularly useful for reservoir sediment management studies. These capabilities have improved the efficiency and accuracy of USACE sustainable sediment management studies and made long term alternative analyses possible.

ACKNOWLEDGMENTS

Features described in this paper were funded by the Corps of Engineers Flood and Coastal Storm Damage Reduction R&D Program and the Missouri River Recovery Project. Dr. John

Shelley at the USACE Kansas City District also provided invaluable insight and support in this process.

REFERENCES

Atkinson,E. 1996. The feasibility of flushing sediment from reservoirs. H.R. Wallingford Report OD 137, London.

Beven, K. (1993), Prophecy, reality and uncertainty in distributed hydrological modeling Advances in Water Resources, 16, 41–51.

Boyd, P.M. & S. Gibson. 2013. Regional Sediment Management (RSM) Modeling Tools: Integration of Advanced Sediment Transport Tools into HEC-RAS. Coastal and Hydraulics Engineering Technical Note ERDC/CHL CHETN-XIV-36. Vicksburg, MS: U.S. Army Engineer Research and Development Center, Coastal and Hydraulics Laboratory.

Cantelli, A., C. Paola, & G. Parker. 2004. Experiments on upstream-migrating erosional narrowing and widening of an incisional channel caused by dam removal. Water Resources Research. Vol. 40, 12 pp. DOI: 10.1029/2003 WR002940.

Cui, Y., G. Parker, C. Braudrick, W.E. Dietrich, & B. Cluer. 2006. Dam removal express assessment models (DREAM). Part 1: Model development and validation. Journal of Hydraulic Research. International Association for Hydro-Environment Engineering and Research. Vol. 44, pp. 291–307. DOI:10.1080/00221686.2006.9521683.

Greiman, B.B. & J. Huang. 2006. One-dimensional modeling of incision through reservoir deposits. Proceedings, 8th Federal Interagency Sedimentation Conference. Reno, NV. pp. 491–497.

Gibson, S. 2013a Comparing Depth and Velocity Grids Computed with One- and Two-Dimensional Models at Ecohydraulic Scales, 267p.

Gibson, S. 2013b. The USDA-ARS Bank Stability and Toe Erosion Model (BSTEM) in HEC-RAS. *Advances in Hydrologic Engineering*. Davis: Hydrologic Engineering Center 10p.

Gibson, S. (2011) "Modeling Long Term Impacts of Sediment on the Capacity of Arghandab Reservoir," Chapter 2 of a Report on the Argandab Reservoir by the Omaha District of the U.S. Army Corps of Engineers, 50 p.

Gibson, S., Brunner, G., Piper, S., & Jensen, M. (2006) "Sediment Transport Computations in HEC-RAS." Eighth Federal Interagency Sedimentation Conference (8thFISC), Reno, NV, 57–64.

Langendoen, E.J., & Simon, A. 2008. "Modeling the evolution of incised streams. II: Streambank erosion." J. Hydraul. Eng., 134(7).

Morris, G.L., Annandale, G., & Hotchkiss, R. 2007. Reservoir sedimentation. In M. Garcia (ed) *Sedimentation engineering: processes, measurements, modeling and practice*. ASCE Manual of Practice 110:570–612.

Morris, G.L. and Fan, J. 1998. *Reservoir sedimentation handbook.* New York: McGraw Hill.

Schumm, S.A., Harvey, M.D. & Watson, C.C. 1984. Incised Channels: Morphology, Dynamics, and Control. Water Resources Publications: Littleton, 200 p.

Simon, A., Curini, A., Darby, S.E., & Langendoen, E.J. (2000). "Bank and near-bank processes in an incised channel", Geomorphology, 35, pp. 193–217.

USACE. 2013. Lewis and Clark Lake Sediment Management Study—Phase II Fact Sheet. Omaha District.

Reservoir Sedimentation – Schleiss et al. (Eds)
© *2014 Taylor & Francis Group, London, ISBN 978-1-138-02675-9*

Global analysis of the sedimentation volume on Portuguese reservoirs

F. Taveira-Pinto, L. Lameiro, A. Moreira, E. Carvalho & N. Figueiredo
Faculty of Engineering, University of Porto, Porto, Portugal

ABSTRACT: This paper aims to examine the process of reservoir sedimentation, namely to try to estimate the sediment volume deposited in the main Portuguese reservoirs. The changes in fluvial sediment transport induced by the construction of dams on the major Portuguese rivers are described. Based on the values of dead volumes, the current situation of 166 Portuguese reservoirs is analysed to obtain an order of magnitude of available sediments and pre-select reservoirs that could potentially be integrated into a project for the artificial sand nourishment of beaches. This work was carried out as a starting point in order to implement a national plan for the use of reservoir sediments as an "added value" to the economy, which is as yet generally unexploited.

1 INTRODUCTION

Over recent decades, dams and reservoirs have been built with increasing frequency for various purposes such as hydroelectricity production, water supply for irrigation, fire protection needs or flood control. Despite the many gains realized from their construction, they also bring environmental impacts as they represent an obstacle for the natural processes of sediment transport in a river. A number of sedimentation issues, namely the loss of volume storage in reservoirs and upstream retention of sediments that would nourish eroding beaches are raising concerns about sediment management policies.

Changes in the flux of sediments from rivers to oceans owing to worldwide construction of hydroelectric dams and reservoirs for irrigation have been reported by many authors (Miliman, 1990; and Poulos & Chronis, 1997). At present it is widely accepted that sedimentation in reservoirs and coastal erosion are co-related problems. It is estimated that around 40,000 large reservoirs suffer from sedimentation worldwide, and between 0.5% and 1% of total storage capacity is lost per year (White, 2001).

Several methods have been developed to forecast the sedimentation in reservoirs, including empirical and numerical models (see Mamede, 2008 for a detailed review). Siltation is generally undesirable and unavoidable, so sediment management strategies are essential to achieving sustainable use of reservoirs. Case studies on the accumulation of sediments in reservoirs and the application of engineering techniques to mitigate it have been done worldwide, for example in Japan (Kantoush et al., 2011), Iran (Boroujeni, 2012), and Australia (James & Chanson, 1999, Lewis et al., 2013).

Methodologies to manage sediments in reservoirs comprise the reduction of sediment yield with measures in the catchment area (e.g. reforestation), sediment routing, sediment flushing; and sediment removal by mechanical dredging (Morris & Fan, 1998, Batuca & Jordan, 2000). The feasibility of flushing sediments from reservoirs through sluices has been reported by Paul & Dhillon (1988), with beneficial results for both small and large capacity reservoirs, describing examples in the USSR and China. The same authors also present guidelines to calculate the optimal dimensions of the sluices.

Even if reduced accumulation or removal of sediments is technically possible, the economic viability is likely to depend on physical, hydrological and financial parameters (Palmieri et al.,

2001). The selection of a single solution or a combined approach is always subject to an economic balance between costs and benefits. A substantial increase in reservoir efficiency along with the benefit of exploiting the extracted sediments in a valuable application can help to justify such investments. Possible exploitations of sediments extracted from silted reservoirs include construction and industrial uses, fertilization of soils (see Fonseca et al., 2003) and artificial beach nourishment.

In the particular case of Portugal, the reduction of river sediment transport due to damming is recognized as a cause of aggravating coastal erosion (e.g. Dias, 1993, Silva et al., 2007, Taveia-Pinto et al., 2011). Hence, the possibility of using dam reservoir sediment deposits for beach nourishment is of interest. In a project for reusing these sediments to nourish beaches some aspects must be considered, particularly; reservoir sediment availability, extraction methods, treatment (grain size separation and decontamination), and transportation.

In this paper the current levels of sedimentation in 151 (out of 166) Portuguese reservoirs were estimated on the basis of the dead volumes defined at the design stage. This is a feasible bottom line for a national program of exploitation of reservoir dead volumes, which can have an added value to the Portuguese economy. Considerations for the selection of reservoirs with the potential to provide sediments for artificial nourishments are also described.

This issue represents a type of investigation that is likely to increase in Portugal and abroad, as the need for environmentally sound solutions for coastal erosion grows.

2 PROCESSES OF SEDIMENTATION IN A RESERVOIR

The construction of a dam modifies sediment transport conditions, usually representing a breach in the normal sequence of erosion, transport and deposition, with the formation of deltas in the upstream border of the reservoir, silting in of the border and erosion following the restitution, Figure 1, (Lysne et al., 2003).

Therefore, when designing a reservoir it is mandatory to provide space for sediment deposition in the dead volume, which is defined according to a scheme of the river transportation of solids and the probable working life of the structure (Lencastre and Franco, 2006).

The deposition of the biggest part of the sediments occurs upstream of the reservoir, in the useful storage volume and not in the dead volume (volume below the lowest level of the water intake). After deposition, the coarse materials show a much greater stability, maintaining their characteristics over time and with a more difficult motion. On the contrary, soft sediment deposits have an evolution of their properties, with compression over time allowing the superficial layers to be placed in suspension much easier.

3 REUSE METHODOLOGIES

The sediments on the bottom of reservoirs are known as preferential placement of organic deposits and mineral materials, namely nutrients, heavy metals and bacteria, and are generally considered as sources of pollution.

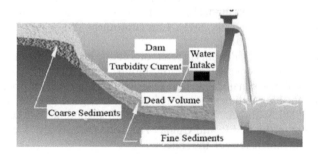

Figure 1. Typical pattern of sediment deposition in reservoirs.
(http://www.dha.lnec.pt/nre/english/projects/sedim_eng.html).

Aiming at artificial beach nourishment, it will be important to evaluate the heavy metal concentration in the sediments, however the organic matter content is more important as a fertile growing medium for plants to stabilize the dunes, providing good conditions for the development of vegetation. The treatment process of sediments removed from reservoirs in order to prepare them for various purposes may be divided in two different stages: granularity distribution and decontamination.

The granularity distribution is usually obtained through bolting. For beach nourishment the required fraction corresponds to the sand fraction (dimensions between 0.075 mm, #200 and 4.75 mm, #4), the granularity should be as close as possible to that of the natural sediments of the beach to be nourished. Beaches nourished with higher granularities will be more stable but steeper, inducing new profiles with possible implications on the comfort and beach safety level. Beaches nourished with smaller granularities are potentially more unstable and so the volumes placed will tend to be washed away sooner by the sea.

It has been accepted that pollution particles are concentrated in the softer fraction of the sediments (Colandini, 1997, Lee et al., 1997, Zanders, 2005—referred to by Pétavy et al., 2008). Nevertheless, recent studies show that the metals don't establish themselves in a specific fraction of sediments, (Durand, 2003, Clozel et al., 2007—referred to by Pétavy et al., 2008).

Therefore, it is always expedient to check the need for decontamination through chemical analysis before proceeding with the transportation of the sediments to apply on beaches. The need for sand decontamination incurs expensive costs in a sediment reuse project for artificial beach nourishment, which can lead to economic unfeasibility and so it should be carefully evaluated. Nevertheless, this process of decontamination may normally be dismissed in regard to the coarser fraction.).

4 APPROACH AND ANALYSIS OF THE PORTUGUESE CASE

4.1 Introduction

The natural erosion phenomenon has become a problem of growing intensity due to human intervention in coastal areas since low altitude areas that were capable of naturally adapting themselves to the variations of average sea levels, storms and tides have now lost much of this capacity, due to the occupation of the coastline with roads, urbanization, playgrounds and industrial areas, among other infrastructures.

The construction of dams has been identified as one of the main causes for the reduction of sediment transportation in rivers, and it is estimated that this reduction reaches 80% in Portugal, (Fonseca, 2002). This contributes significantly to the need for engineered interventions and in 2008, the related investment in coastal defence works was 7,579,000€ (Maia, 2009).

4.2 Approach to the Portuguese case

In Portugal, large forest fires occur quite often during the summer, destroying the vegetation over large areas and increasing the affluence of sediments to watercourses and finally to the sea, where the main direction of sediment transport along the Atlantic coast is from north to south. In order to give an idea on the capacity of sea transport some references indicate that it was in average nearly 1,000,000 m^3/year from the Douro River mouth to Nazaré (Veloso-Gomes et al., 2006b).

In 1930, prior to the construction of dams, the affluence of sediments to the Portuguese North-West shore was approximately 2,000,000 m^3/year, with the River Douro itself assuring 90% of this value (PNA, 2008). Nowadays, as a result of dam construction that has reduced the flow speed and subsequent ability to erode the river bed, the Douro River only contributes a small amount of sediment, estimated at 250,000 m^3/year.

In addition, there is also a significant amount of sand extracted for the construction industry, usually dredged from the river bed in places and in amounts beyond what it would be sustainable (Veloso-Gomes et al., 2006a).

The mean rates of sedimentation calculated in the period between 1983 and 1984 in the Tejo estuary, indicated values of 1,1 and 1,5 cm/year. During the construction of the Vasco da Gama Bridge about a million m³ of sediment was dredged from the Tejo River, which was contaminated in some way (PNA, 2008). In the Guadiana River one of the consequences of flow reduction induced by hydrological loss and regularization of the huge installed storage capacity (dominated 70,4% of the basin in 1990) is decreased sediment supply, also for the adjacent coastal areas (PNA, 2008).

In Portugal the importance of the evaluation of the reservoir dredging is recognized, as well as its consequences and the need for a permanent group for its analysis.

Since 1996 the Water Institute (INAG) has been restructuring the monitoring networks of hydraulic resources, including the sediment network. The new sediment network includes stations in watercourses where samplings of solid flows will be performed, bottom granularities and stations in reservoirs where bathymetric surveys will be performed. As the network is still in implementation the results are not yet known.

4.3 Balance and analysis of the current sedimentation in the main Portuguese reservoirs

Despite the efforts undertaken to overcome the lack of information regarding sediment transport in Portugal the amount of deposited sediments in the reservoirs still remains unknown, and consequently so does the actual capacity of the Portuguese reservoirs.

Therefore, the data that supports this study comes from the inventory of Portuguese dams made by INAG. The dams in this inventory generally comply with the criteria of being more than 15 m in height or having more than a million m³ of total storage capacity.

Using the values presented by INAG for dead-volumes and the information regarding the siltation rate (Lencastre and Franco, 2006) an attempt was made to estimate values in a way that can show the current situation of the sedimentation in these Portuguese dams.

In this estimation the designed dead volume of each dam was considered as the real volume of actual sedimentation in a reservoir with 50 or more years of service.

$$\text{Volume}_{sedimentation} = \text{Volume}_{dead} \tag{1}$$

and for dams with less than 50 years of service it was defined that the sedimentation of the reservoir would be directly proportional to the age of the dam:

$$\text{Volume}_{sedimentation} = \frac{\text{Volume}_{dead}}{50} \cdot \text{age}_{dam} \tag{2}$$

With the purpose of being able to bridge the lack of data, the procedure was similar to the suggestion for the volume balance of the deposited sediments volume referred to in Basson (1999), in which an empirical relation between the basin area and the volume of deposited sediments is established.

This correlation was defined separately for the hydrographic areas of the North, Center and South in an attempt to isolate cases with similar conditions of erosion and transportation, Figures 2–4.

Although these relationships do not present high degrees of linear correlation, it was considered that they are capable of supplying sediment volumes with enough accuracy.

It is necessary to highlight that base assumptions were adopted in order to obtain an order of magnitude, and that once more accurate and detailed data is available other analysis could be done.

The INAG database consists of 166 dams and it was possible to gather enough information to estimate the volume of deposited sediments in 151 of the 166 dams.

It is estimated that these 151 dams have a global volume of deposited sediments of 1,568 million m³ in a total gross capacity of 12,546 million m³, which means that 12.5% of the total gross capacity for the water storage is now occupied by sediments.

Of the 151 dams with estimated sedimentation volumes, about 66% have a sedimentation volume below 10% of their gross storage capacity, 28% are between 10% and 50% and only 6% are above 50% of gross storage capacity.

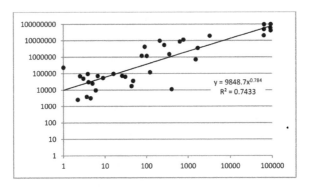

Figure 2. Relationship between the deposited sediments volume (yy-m³) and the hydrographic basin area (xx-km²) in the North region of Portugal.

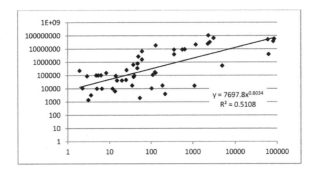

Figure 3. Relationship between the deposited sediments volume (yy-m³) and the hydrographic basin area (xx-km²) in the Center region of Portugal.

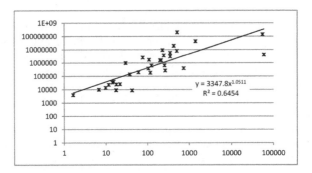

Figure 4. Relationship between the deposited sediments volume (yy-m³) and the hydrographic basin area (xx-km²) in the South region of Portugal.

Using ArcGis software, the coordinates of the dams were referenced in the Geographic Information System, which allowed an analysis of the positioning of the 166 reservoirs in regard to the Portuguese coast, Figure 5. Portugal is divided into five Hydrographic Regions that are also defined in Figure 5.

This division is the most adequate, since observation of the map allows for faster reading of the reservoirs' distribution through the Hydrographic Regions that have jurisdiction over the correspondent coastal areas, suggesting a way of managing the entire process for reuse of the sediments.

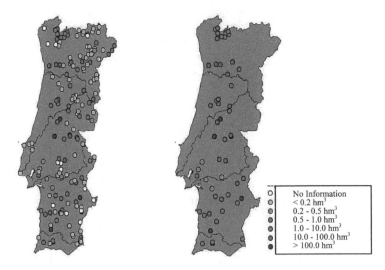

Figure 5. Localization of the 166 reservoirs considered (left), and selected (right), INAG.

The reservoirs that would be more interesting for a project of reusing the sediments for beach nourishment were selected. The selection criteria were the following:

– Reservoirs with over than 0.2 hm³ of sediments and located less than 50 km from the coast;
– Reservoirs located more than 50 km and less than 100 km from the coast and with more than 0.5 hm³ of sediments.

These criteria are similar to those of a study done in Spain for the same purpose (Sanchez, 2008). The proposed criteria are aimed at supporting reservoir pre-selection, and so further investigation is required to come to a final conclusion. Of the 166 reservoirs analysed 53 were selected, representing 1,242 million m³ of sediments and 10,968 million m³ of gross capacity for water storage.

Of the reservoirs selected, 11 are located less than 15 km from the coast, 17 are between 25 and 50 km, 10 between 50 and 75 km and 15 between 75 and 100 km, respectively representing 65, 484, 345 and 347 million m³ of available sediment, Figure 5.

As regards the hydrographic regions, the pre-selection comprises 269 million m³ in the North Hydrographic Region, 84 million m³ in the Center and 482 million m³ in the Tejo Hydrographic Region, 394 million m³ in the Alentejo and 13 million m³ in the Algarve.

The Hydrographic Regions of Tejo and Alentejo are those that present the largest amount of available sediments, perhaps due to the presence of large dams. The Algarve and Central hydrographic regions present a smaller amount of available sediments, and they are also the regions with fewer reservoirs analysed.

5 FINAL REMARKS

In Portugal, there is currently no strategy for the management of sediments, although the sedimentation problem in reservoirs is recognized along with the problem of coastal erosion. It appears with less intensity as it is of small concern to society in general, but in the case of no intervention it will become a problem to be faced in the future.

So, the sooner we plan a strategy for intervention, the smaller will be the effort required to implement it. Also, there is no single solution for this problem, as each reservoir has its own specific characteristics.

Transportation is the conditioning factor of a project for reuse of sediments for beach nourishment. Transport by land is expensive and has significant negative impacts that increase with distance. This requires a search for alternative solutions. Transport by river can be a solution in the navigable sections of the main rivers.

A cost-benefit analysis should take into account the benefits from not only the reuse of sediments but also the reduction of high costs associated with coastal defence works and the maximization of profit from the dams, which, from the major impacts they represent, justify the maintenance work that will allow their proper function.

From all this it is important to recognize that the current investment in the revitalization of the sediment networks constitutes an excellent opportunity to define an effective management strategy with as extensive as possible use of the sediments, which could be a major benefit to the national economy.

Such projects require technical studies, financial support and political will. The coordination of all parties involved is mandatory: water companies, industrial promoters, coastal and environmental authorities and other users: residents, fishermen and land owners surrounding the reservoirs.

This work was done as a starting point, in order to implement a national plan for the exploitation of sediments, which are considered to be an important benefit to the economy but are generally yet unexploited.

REFERENCES

Álvares, M.T., Fernandes, S., Mariano, A.C. & Pimenta, M.T. 2000. Monitorização batimétrica para Gestão de Albufeiras: Estudo Piloto. IX SILUBESA, 2000, Porto Seguro—Brasil.

Álvares, M.T., Fernandes, S., Mariano, A.C. & Veríssimo, M.R. 2001. Plano de Trabalhos para Execução de Levantamentos Batimétricos nas Albufeiras da Rede Sedimentológica. Instituto da Água—Direção de Serviços de Recursos Hídricos.

Basson, G.R. & Rooseboom, A. 1999. Dealing with reservoir sedimentation, Comissão Internacional das Grandes Barragens (ICOLD), Paris.

Batuca, D.G. & Jordaan, J.M. (Jr) 2000. Silting and Desilting of Reservoirs. A.A. Balkema, Roterdão.

Boroujeni, H.S. 2012. Sediment Management in Hydropower Dam (Case Study—Dez Dam project), HydroPower—Practice and Application, Dr. Hossein Samadi-Boroujenin (Ed.), ISBN: 978–953–51–0164–2, InTech, DOI: 10.5772/33115.

Dias, J.M.A. 1993. Estudo da Avaliação da Situação Ambiental e Proposta de Medidas de Salvaguarda para a Faixa Costeira Portuguesa. Geologiacosteira. Chapter IV, pp 13–38 (in Portuguese).

Fonseca, R.M.F. 2002. Impactos ambientais associados a barragens e a albufeiras. Estratégia de reaproveitamento dos sedimentos depositados. CGE/Universidade de Évora (in Portuguese).

Fonseca, R., Barriga, F.J.S.A. & Fyfe, W.S. 2003. Dam Reservoir Sediments as Fertilizers and Artificial Soils. Case Studies from Portugal and Brazil. Proceedings, Water and Soil Environments, Biological and Geological Perspectives, KazueTazaki (Ed), International Symposium of the Kanazawa University, Vol 21. 55–62.

http://sites.google.com/site/geologiaebiolobia/_/rsrc/1219775711387/Home/geologia-problemas-e-materiais-doquotidiano/bacias-hidrogr%C3%A1ficas (30/01/2009).

James, D.P. & Chanson, Hu. 1999. A Study of Extreme Reservoir siltation in Australia. Water 99 Joint Congress, 25th Hydrology and Water Resources Symposium and 2nd International Conference on Water Resources and Environmental Research, Brisbane, Australia, (987–992), 6–8 July.

Kantoush, S.A., Sumi, T. & Takemon, Y. 2011. Lighten the Load. International Water Power and Dam Construction, May, 38–45.

Lencastre, A. & Franco, F.M. 2006. Lições de Hidrologia, Fundação da Faculdade de Ciências e Tecnologia da Universidade Nova de Lisboa, Caparica (in Portuguese).

Lewis, S.E., Bainbridge, Z.T., Kuhnert, P.M., Sherman, B.S., Henderson, B., Dougall, C., Cooper, M. & Brodie, J.E. 2013. Calculating Sediment Trapping Efficiencies for Reservoirs in Tropical Settings: A case Study from the Burdekin Falls Dam, NE Australia. Water Resources Research, Vol. 49, 1017–2029.

Lysne, D., Glover, B., Støle, H. & Tesaker, E. 2003. Sediment Transport and Sediment Handling, Hydraulic Design. Norwegian University of Science and Technology Department of Hydraulic and Environmental Engineering, Trondheim, Norway, 117–155.

Maia, A. 2009. Recuperação de Praias Custou 18 Milhões em 2008. Jornal de Notícias, 21/04/2009, 30 (in Portuguese).

Mamede, G.L. 2008. Reservoir Sedimentation in Dryland Catchements: Modeling and Management. PhD Thesis, University of Postdam, Germany.

Milliman, J.D. 1990. Fluvial Sediment in Coastal Seas: Flux and Fate. Unesco Parthenon Publishing. Nature and Resources, 26,4: 12–22.

Morris, G.L. & Fan, J. 1998. Reservoir Sedimentation Handbook: Design and Management of Dams, Reservoirs and Watersheds for Sustainable Use. New York: Mc Graw Hill.

Palmieri, A., Shah, F. & Dinar, R. 2001. Economics of Reservoir Sedimentation and Sustainable Management of Dams. Journal of Environmental Management, 61, 149–163.

Paul, T.C. & Dhillon, G.S. 1988. Sluice Dimensioning for Desilting Reservoirs. Water Power and Dam Construction, May.

Pétavy, F., Ruban, V. & Conil, P. 2008. Treatment of StormWater Sediments: Efficiency of an Attrition Scrubber—Laboratory and Pilot-scale Studies. Chemical Engineering Journal, 28/04/2008, página 477, Elsevier B.V.

PNA—Plano Nacional da Água. Capítulos 5, 7, 8 e 10 do II Volume.

Poulos, S.E. & Chronis, G. TH. 1997. The Importance of the River Systems in the Evolution of the Greek Coastline. Bulletin de l'Institute océanographique, Monaco, n° special 18, pp 75–96.

Sánchez V. 2008. Sedimentation of the Spanish Reservoirs as Sand Source for Beach Nourishment. Área de Estudos de Costas, Centro de Estudos de Portos e Costas (CEDEX), Madrid—Espanha.

Silva, R., Coelho, C., Veloso-Gomes, F. & Taveira-Pinto, F. 2007. Dynamic Numerical Simulation of Medium Term Coastal Evolution of the West Coast of Portugal. Journal of Coastal Research, SI 50 (Proceedings of the 9th International Coastal Symposium), pp 263–267, Gold Coast, Australia.

Taveira-Pinto, F., Silva, R. & Pais-Barbosa, J. 2011. Coastal Erosion Along the Portuguese Nortwest Coast Due to Changing Sediment Discharges from Rivers and Climate Change. Global Change and Baltic Coastal Zones, Coastal research library, Volume 1, pp 135–151.

Veloso-Gomes, F., Taveira-Pinto, F, das Neves, L., Pais-Barbosa, J. 2006a. O Projecto EUrosion. Resultados e Recomendações para uma Gestão mais Eficaz da Erosão Costeira. 8° Congresso da Água (FCTUC), 13/03/2006, Figueira da Foz, páginas 1–9, IHRH-FEUP, Porto (in Portuguese).

Veloso-Gomes, F., Taveira-Pinto, F., das Neves, L., Pais-Barbosa, J. 2006b. EUrosion—Pilot Site ofRiver Douro—Cape Mondego and Case Studies of Estela, Aveiro, Caparica, Vale do Lobo and Azores. Faculdade de Engenharia da Universidade do Porto e Instituto de Hidráulica e Recursos Hídricos.

White, R. 2001. Evacuation of Sediments from Reservoirs. Thomas Telford, London.

Reservoir Sedimentation – Schleiss et al. (Eds)
© *2014 Taylor & Francis Group, London, ISBN 978-1-138-02675-9*

Reservoir sedimentation management at Gebidem Dam (Switzerland)

T. Meile
Basler & Hofmann, Zollikofen, Switzerland

N.-V. Bretz
HYDRO Exploitation, Sion, Switzerland

B. Imboden
Electra-Massa AG, Sion, Switzerland

J.-L. Boillat
EPFL, Lausanne, Switzerland

ABSTRACT: The Gebidem Dam located on the Massa River intercepts annually around 400'000 m³ of solid material. These sediments must be evacuated by annual flushing of the reservoir over 4 to 7 days. A trend for silting in the flushing channel was observed during the last decades, leading to damageable overtopping of the lateral walls. In order to improve knowledge on input and output of sediments in the channel of Gebidem, physical and numerical modeling was performed in 1994 and in 2002 at the Laboratory of Hydraulic Constructions of the École Polytechnique Fédérale de Lausanne (EPFL). It was clearly demonstrated that the silting process is related to an anti-dunes regime progressing in the upstream direction from deposit zones which develop initially in the curves and the river confluence with Upper Rhone River. At the opposite, the clearing process starts at the upper limit of the channel and progresses by pushing the sediments downstream. The proposed solution requires an additional water supply to be introduced from the entrance of the channel. In order to optimize the clearing efficiency, the dilution supply can be progressively reduced as soon as the sediment concentration diminishes. Being aware of the importance of the annual flushing, which is the only measure to guarantee sustainable operation of the hydropower scheme, Electra-Massa continuously undertakes efforts to ensure and further improve flushing operations.

1 INTRODUCTION

1.1 *General remarks*

Dams significantly influence the sediment balance on watersheds. A specific management is therefore necessary to preserve normal operating conditions of the stored water as well as to keep the usable volume of the reservoir. Most of Swiss dams were built in the middle of last century and the dead zone assigned to sediment storage is generally full, leading to operation difficulties at intakes and bottom outlets (Boillat et al., 2000 and 2003). Considering that main part of the Swiss electrical production issues from hydropower schemes, it becomes obvious why reservoir sedimentation management is a major challenge.

Different solutions exist to intervene in sediment transport and deposition processes, generally related to operational and flood safety as well as to sustain environment. Complementary actions can be carried out upstream in the watershed, in the reservoir itself, at particular locations near the dam and downstream as well (Fig. 1). Among them, sediment flushing reveals an efficient solution to evacuate sediments from a reservoir, generally by opening the

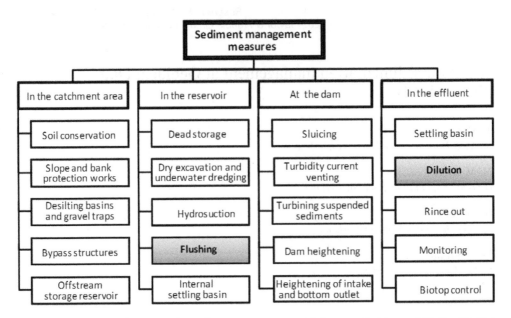

Figure 1. Inventory of measures for sediment management with focus on flushing and dilution (Boillat et al., 2003).

bottom outlet. This procedure, which is yearly applied to the reservoir sediment management at Gebidem Dam, requires a careful planning in order to avoid material and ecological damage to the downstream river reach.

Swiss legislation about reservoir management concerns two objectives, the safety of dams on the one hand and the water and fish protection on the other hand (Boillat and Pougatsch, 2000, Pougatsch et al., 2002). The safety issue is governed by the Federal Law regarding Supervision of Hydraulic Structures. Besides, the general aims of the Swiss water protection law are to avoid consequences on human health, animals and vegetation. As a consequence, the owner of a dam shall ensure as far as possible that flushing or emptying of a storage basin does not adversely affect the fauna and flora in the river downstream. Furthermore, periodical flushing requires a cantonal authorization fixing the time and duration, the maximum suspended load concentration and the conditions of post-flushing operations to clear out the riverbed.

1.2 Massa-River catchment

The Massa-River catchment area is 198 km² at Gebidem Dam (1436 m a.s.l.) and culminates at 4191 m a.s.l. The watershed counts several glaciers (Fig. 2) with a total cover of 63.9% leading to a glacier and snowmelt dominated flow regime with 83% of the yearly runoff during only 4 months, from June to September. The mean annual runoff was estimated at about 429 Mio m³ over the period 1981 to 2000, with a mean discharge of 13.62 m³/s.

Considering the usable storage volume of 5.8 Mio m³ at Gebidem Dam, 74 filling and emptying cycles are theoretically possible annually. However, due to very high flow rates during summer months, Gebidem Dam is operating as a runoff river hydroelectric scheme during long periods and is acting as a huge sand and silt trap.

Due to strong glacier melting, an increase of the mean annual runoff up to 470 Mio m³ was observed during the last 10 years and will probably culminate at 550 to 560 Mio m³ in 2050 (SGHL and CHy, 2011). After this peak, a rapid decrease to about 450 Mio m³ in 2070 is forecasted, based on Swiss Climate Change Scenarios CH2011 of C2SM, ETHZ (Bosshard et al., 2011).

The runoff carries an important volume of sediments which amounts to about 0.1% or some 430'000 to 470'000 m³ per year since 2001. Around 10% crosses the Gebidem Dam towards the

Figure 2. Left: Aletsch glacier in the Massa-River catchment (www.swisseduc.ch; 1994). Right: Gebidem Dam and lake (J. Germanier, HYDRO Exploitation SA; 2010).

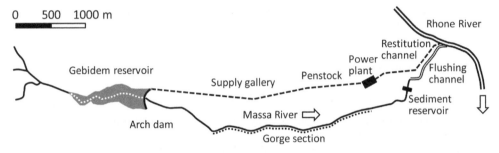

Figure 3. Elements of the hydropower scheme of Electra-Massa. Not part of scheme: sediment reservoir.

hydropower plant (Alpiq, 2010), whereas 90% is retained and must be flushed. The sediments range from blocs/gravel to clay with a mean grain size of 1 to 3 mm (LCH-EPFL, 2002).

The evolution prognosis of the yearly amount of sediments due to climate change in the catchment is however not certain. Many factors such as increase of sediment availability, permafrost melting, discharge increase or floods frequency raising might be balanced by glacier lakes formation, flattened reaches with limited sediment transport or appearance of bedrock.

1.3 *Hydropower scheme of Electra-Massa SA*

The Gebidem Dam was built between 1964 and 1967 across the Massa River with the purpose of hydro-electricity production. The main elements of the hydropower scheme and the intermediate sediment reservoir (not part of the hydropower scheme) are schematized on Figure 3:

- Catchment area 198 km^2
- Reservoir at 1436 m a.s.l. with a total volume of 9.2 Mio m^3 (usable volume 5.8 Mio m^3).
- Arch dam 122 m high, with a crest spillway of 350 m^3/s capacity.
- Supply gallery over 2'685 m length, with a 3.40 m diameter.
- Penstock with 2.50 m diameter and 1100 m length, under 743 m head.
- Power plant with a total installed power of 340 MW, shared in 3 units equipped with 55 m^3/s Pelton turbines.
- Flushing channel with 1.5% longitudinal slope and 700 m length, 8 to 10 m width and 5 to 8 m depth.
- Intermediate sediment reservoir upstream of the flushing channel, with a maximal retention volume of about 30'000 m^3, behind a concrete wall of about 30 m height.

1.4 *Goals and importance of secure and sustainable flushing*

Without flushing operations, the total storage volume of Gebidem Dam (9.2 Mio m³) would be completely filled in 20 to 25 years. After 6 to 7 years without flushing, the level of the intake to the headrace tunnel would be attended by the alluvium level, leading to unacceptable high sediment load in the power plant. Already sooner, safe use of the bottom outlet could be uncertain, with some structural concerns appearing as the dam was not designed for sediment pressure. Thus, a yearly flushing must be carried out for sustainable use of the hydropower scheme.

The flushing operation is particularly demanding for the following reasons: 1. Very high amount of sediments. 2. Only few days or weeks exist during May and June with acceptable discharge conditions in the Massa and the Upper Rhone rivers. 3. Challenging general setup of the longitudinal profile with a slope reduction between the Massa River (4–5% in sediment transport limiting reaches) and the flushing channel (1.5%). 4. Ecological, economic and industrial constraints related to legally authorized suspended load, limited loss of water for energy production and use of the Upper Rhone River for cooling processes in the industry.

The goal of the yearly flushing can thus be summarized as: "Sustainable and secure for men, environment and infrastructures". Where sustainable in the flushing context means: "All sediments are yearly removed from the Gebidem Dam, the Massa River and Gorge as well as the flushing channel."

2 THE ANNUAL FLUSHING

2.1 *Processes in reservoir, Massa River and Gorge, flushing channel and Upper Rhone River*

Many different processes related with river hydraulics and sediment transport are observed during the annual flushing (Fig. 4).

Since the reservoir acts as huge sand and silt trap, finer sediments are settled down next to the dam while coarse material is accumulated in the delta which develops during summer. When opening the bottom outlet for flushing, fine sediments are first discharged leading to an abrupt increase in sediment concentration. The reservoir is then totally emptied, leading to incision of the channel bed with lateral erosion and sliding of hillside deposited material, which temporarily increase the sediment concentration. Coarser material, transported as bed load, and finer sediments in suspension pass the Gebidem Dam simultaneously.

In the Massa-River between the bottom outlet (ca. 1'328 m a.s.l.) and the entrance of the flushing channel (692.55 m a.s.l.), the bed load transport takes around 60 to 72 hours to cover the Gorge while suspended load runs near as fast as water.

Figure 4.   Flushing operation 2003. Left: Erosion and transport process in the reservoir after complete drawdown. Right: Fluid-solid mixture downstream from the bottom outlet during a flushing operation.

Thus, transport processes in the 700 m long flushing channel linking the Massa-Gorge with the Rhone River are dominated during two to three days by a highly saturated flow, poor in bedload. After two or three days, a trend of deposition is observed in the channel due to a lack of sediment transport capacity. The silting process progresses from downstream to upstream, first initiated at channel bends as well as at a channel contraction and at the Rhone River confluence.

At the rivers junction, the highly concentrated Massa flow is mixed to the Rhône water over several hundred meters leading progressively to a reduction of the suspended load concentration. Bed load sediment transport at the junction depends on the total flow of both rivers. Nevertheless, an increase of the bed level is commonly observed on the right half of the Rhone River downstream from the confluence.

During emptying and clearing operations of reservoirs, according to the cantonal guidelines, a maximum sediment concentration of 10 ml/l has to be respected. When applying this rule, 40 mio. m³ water would be necessary to evacuate 400'000 m³ of sediments. This water volume represents 10–16% of the annual contribution. The actual flushing operations at Gebidem present sediment concentrations 4 to 6 time higher than the required value. Such a condition is admitted in this particular case, because the rocky canyon and the concrete channel downstream of the dam do not accommodate any particular natural life.

2.2 Yearly flushing operation

The sediment amounts which have been removed from the reservoir during the yearly flushing have been systematically evaluated since 1969 (Fig. 5). Datasets before 1991 must be considered with care as the sediment amount was directly deduced from the volume of used water and not from photogrammetric or bathymetric measurements. Furthermore, some data (1979, 1980, 2008, 2009, 2013) are missing.

Considering the last twenty years, an amount of 300'000 to 400'000 m³ needed to be flushed yearly during the appropriate period, normally from May to June. Flushing in this season is appropriate regarding the respective discharges of the Massa and Rhone rivers, and also environmental reasons. In fact, the flow increases at the end of the winter period in both rivers and succeeding floods due to strong snow and glacier melt occur at the second half of June. These floods help to restore the river bed and to evacuate the remaining fine sediment deposits.

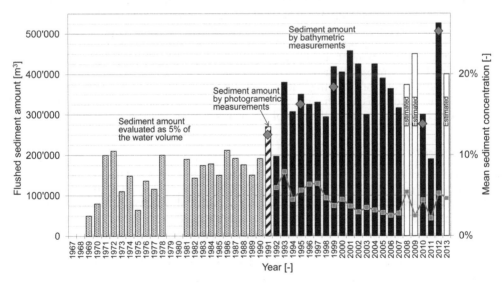

Figure 5.   Yearly amount of flushed sediments and mean sediment concentration (since 1992) at Gebidem Dam (-■-). Critical flushing operations: ◆.

249

Figure 6. Left: Surface waves due to anti-dunes regime during sedimentation in the flushing channel. Right: High sediment load leading to channel overtopping in 1999.

As mentioned, the yearly flushing is a challenging operation. Several times, difficult situations appeared as documented for years 1991 and 1993, with external costs of more than 400'000.- Swiss francs (EOS, 1999). The most critical flushing operation occurred in 1999 (EOS, 2002), when permanent and competent monitoring and field operations were required to limit damages as far as possible (Fig. 6). Similar problems also occurred in 2010, for reasons still under clarification, and in 2012 with very high sediment amount and inopportune failure of a sluice gate in the intermediate sediment reservoir.

3 CONTINUOUS EFFORTS SINCE THE ORIGIN OF THE HYDROPOWER SCHEME

3.1 *Physical modelling in 1964*

Four physical models were operated at the Istituto di Idraulica dell'Università di Padova (Italy), two for the turbine pits, one for the restitution channel and one for the flushing channel. The tests conducted on the last model led to the actual shape of the channel and its 1.5% longitudinal slope. Lower slopes of 0.8% and 1.2% were also investigated, but they revealed insufficient to adequately manage the sediment transport. The study recommended proceeding to an annual flushing operation during 4–5 days with a discharge of 20 m³/s assuming to evacuate about 100'000 m³ of sediments by bedload transport and an additional amount of 400'000 m³ being supposed transported in suspension.

From today's point of view, the study of 1964 identified the main difficulties namely the obligation to build a flushing channel. It also led to an optimization of transport capacity in the section but underestimated the part of sediments transported as bed load.

3.2 *Further tests in 1994 and 2002*

After the difficulties encountered during flushing in 1991 and 1993 and later in 1999, physical and numerical modeling were performed in 1994 (Boillat et al., 1996) and in 2002 (LCH-EPFL 2002) at the Laboratory of Hydraulic Constructions of the École Polytechnique Fédérale de Lausanne. The main purpose was to improve knowledge on relevant sediment silting and transport processes. The first study focused on the upper part of the channel, over a distance of about 400 m. The idea was to produce an initial acceleration of the flow by insertion of a sill at the upstream end of the channel. Different alternatives with bottom deflectors and channel width reduction were simulated in search of an optimal solution. Despite some differences between the tested variants, no efficient design could be obtained (Boillat et al., 1996).

For the last study in 2002, a 1:18 down scaled physical model was built, reproducing part of the Massa gorge and the full reach of the flushing channel over a distance of about 700 m,

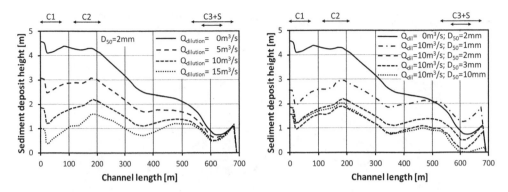

Figure 7. Maximum deposit profile in the channel related to the flushing operation of 1999. Left: Sensitivity analysis of additional clear water supply from 5 to 15 m³/s, for a characteristic grain size $d_{50} = 2$ mm. Right: Sensitivity analysis of the characteristic grain size d_{50} from 1 to 10 mm, for an 1 clear water supply of 10 m³/s. C1, C2, C3 refer to the 3 curves where sediment deposit is initiated. S indicates the location of a shrinkage section in the channel.

up to the Rhone River (Boillat et al., 2003). Two series of tests were conducted, orientated successively towards sedimentation and erosion phases, with the aim to describe and to quantify the respective processes during flushing operations.

It could be put in evidence that silting results from an anti-dunes regime progressing towards upstream. This process is initiated in the existing curves of the channel, where deposit starts in the inner part before progressively covering the whole section. On the other hand, sediment clearing progress from the upper boundary of the channel, eroding and evacuating deposits under a flow stream effect (LCH-EPFL, 2002).

Based on the experimental results, it became obvious that an additional discharge of clear water was required at the upper limit of the channel to increase the sediment transport capacity. In order to evaluate the influence of this dilution flow, a non-steady numerical model was applied down to the Rhone River. After calibration with reference to experimental results as well as to the documented flushing of the year 1999, a sensitive analysis about sediment grain size distribution and additional dilution discharge (Fig. 7) was carried out.

3.3 The proposed solution

Physical model tests and numerical simulations opened out into recommendations for the flushing procedure. It has been suggested that an additional water supply had to be introduced at the upstream end of the channel, as soon as the sediment layer reaches 1.5 m depth at a control section located about 200 m from the entrance of the channel. This reference corresponds to the place where the sediments depth is the highest.

The dilution discharge has to be increased progressively in order to avoid a rapid accumulation of sediments at the input location, leading to a local sediment accumulation and consequently to the water surface increase. In order to optimize the clearing efficiency, the dilution supply will progressively be reduced as soon as the sediment concentration diminishes.

4 THE DILUTION SUPPLY TUNNEL

4.1 The project

According to the proposed solution, the project consisted to build a water supply tunnel, originated from the neighboring low chute Massaboden hydropower scheme (Fig. 8).

The 506 m long tunnel with 4.5% longitudinal slope has three types of sections adapted to geological conditions with around 2.50 m width and 2.50 m height. This device is able to provide a dilution discharge up to 15 m³/s and thus to manage the flooding risk in the flushing

251

Figure 8.   Schematic view of the dilution supply tunnel project.

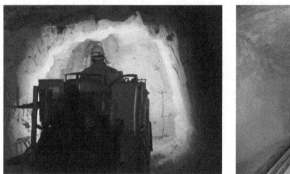 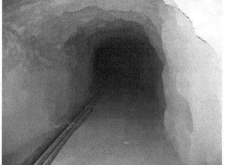

Figure 9.   The dilution supply tunnel under construction (on the left) and completed (on the right).

channel. Furthermore, if considering the reduced time of the flushing operation and the head difference between the Gebidem Dam and the low chute scheme, the investment will be amortized in less than 10 years.

This example of a sediment management measure at the effluent by dilution (Fig. 1) shows the positive contribution of additional water for clearing the deposit and reducing the sediment concentration at the outflow.

4.2   *The construction*

The construction was realized between September 2005 and August 2006. The tunneling method was the traditional drill-and-blast, adequate for such a small section (Fig. 9). First blast was made on 16.11.2005 and the last one on 18.07.2006. A short access tunnel

Year	Water volume Massa River m³	Water volume Supply shaft m³	Sediment Volume m³	Energy Consumption MWh	Energy per Sed. volume MWh/m³
2004	13'400'000	0	425'000	22'110	0.0520
2005	13'600'000	0	390'000	22'440	0.0575
2006	14'400'000	0	364'000	23'760	0.0653
2007	11'370'000	1'860'000	316'000	18'928	0.0599
2008	6'250'000	3'230'000	373'750	10'603	0.0284
2009	12'870'000	5'200'000	450'000	21'704	0.0482
2010	5'040'000	1'800'000	300'000	8'478	0.0283
2011	7'440'000	1'320'000	190'000	12'395	0.0652
2012	6'220'000	31'816'000	526'000	13'126	0.0250
2013	5'563'000	3'086'000	400'000	9'457	0.0236

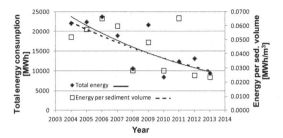

Figure 10. Water supply and evacuated sediment volume during flushing operations, before and after set up of the dilution system in 2007, as well as related energy consumption.

was first excavated, about 20 m near the lower end of the dilution tunnel in the Massa gorge.

The connection of the dilution supply tunnel to the neighboring scheme was achieved only one year later, due to the delayed completion of the transformation works on this scheme.

4.3 New flushing concept and first experiences

The new flushing concept consists to evacuate the sediments from the Gebidem reservoir rapidly with a large discharge, to create an intermediate stock of sediments in the Massa River between the dam and the Rhone River and to wash the Gorge with a reduced discharge, allowing the filling up of the reservoir for the reactivation of energy production. When the sediments layer reaches a certain depth in the flushing channel, the gate of the dilution supply tunnel is opened and its discharge is controlled in function of the evolution of the sediment level.

The first experience of the new flushing concept was successfully made in 2007. With 11'370'000 m³ of water from the Massa River and 1'860'000 m³ from the neighboring scheme, the dilution supply proved its efficiency. Considering differentiated energy coefficients for Gebidem high head and Massaboden low chute schemes, the total energy consumption as well as the energy per sediment volume unit used for the flushing operation could be estimated. The values obtained before and after introduction of the dilution supply in 2007 show a significant trend towards energy efficiency (Fig. 10).

5 FURTHER EFFORTS AND FUTURE CHALLENGES

Despite the long experience of operating Gebidem draw-down for flushing and additional flexibility offered by the dilution supply tunnel, some residual risk remains due to uncertainties in meteorological forecast, sudden weather changes (temperature, clouds, precipitations) in the high mountain catchment, risk associated to handling and state of the infrastructure of the intermediate sediment reservoir as well as sliding, erosion and transport processes of sediments in the reservoir or other unexpected difficulties.

The owner of the Gebidem Dam, Electra-Massa, pursues thus systematic and continuous efforts to ensure further improvement of flushing operations according to the field of actions summarized in Table 1.

The major challenge in the following decades will be to catch the period for secure and sustainable flushing under tendency of decreasing Upper Rhone discharge (lower glacier cover with sooner snowmelt) and increasing Massa discharge up to the year 2050 (high glacier cover).

Table 1. Systematic pursued efforts to ensure further improvement of flushing.

Field of actions	Action	Goal
Optimization of discharge distribution between Massa and Upper-Rhone rivers	– Precise discharge and weather predictions – Ensure use of neighboring power plant with transfer of 7 m³/s from Massa to Rhone – Best use of bottom outlet at Gebidem Dam – Integration of neighboring storage plants to increase Upper-Rhone discharge	– Best moment for flushing – Dilution of sediment concentration in Upper-Rhone – Avoid system overload – Dilution of sediment concentration in Upper-Rhone
Competent and focused use of flushing infrastructure	– Extensive documentation of yearly flushing – Ensure competences of key persons – Punctual upgrading measurement system	Best practice, continuous learning, sharing and transmission of knowledge
Continuous maintenance of infrastructure	– Renovation flushing channel (2015–2017) – Regular use and maintenance of movable component parts	– Certain use of infrastructure – In time perception of necessity for maintenance
Controlling the residual risk	– Precisely defined locations for intervention – Knowing flow paths of water overtopping – Mobile equipment for goods protection	– Enhance security for men – Reduce damage – Reduce loss of income

6 CONCLUSIONS

Reservoir sedimentation management is an essential task worldwide. In Switzerland, where most dams are dedicated to electricity production, mainly by storage plants, the dead storage, designed to accumulate the sediment deposit is generally completely filled. This is a source of problems regarding the safe operation of turbines and bottom outlet devices.

In the particular case of Gebidem Dam, the power scheme functioning is maintained safe thanks to annual flushing operations. However, considering the high amount on sediments to evacuate, the flushing channel suffers a silting process, leading to unacceptable overtopping of the lateral walls. The introduction of an additional clear water supply at the upper limit of the channel was proposed with the aim to reduce the sediment concentration and to control the settling process in the channel.

This solution was first tested on physical model, providing reference data for the calibration of a numerical model. A sensitivity analysis on the dilution discharge and the characteristic grain size allowed then to define operation rules for flushing.

The conceptual idea could be achieved by water diversion from a neighboring low chute hydropower scheme in 2006. Recent experiences with a dilution water supply confirm the efficiency of this solution. However, residual risk cannot be totally excluded and continuous and systematic efforts need to be undertaken for safe and sustainable flushing in the future. These concern namely the field of actions "optimization of discharge distribution between Massa and Upper-Rhone".

The acquired experience about the Gebidem reservoir sediment management allows tending towards optimal flushing operations. This issue could only be attained thanks to a great perseverance, confirming the absolute necessity to consider the problem of reservoir sedimentation already in the early stage of the design for new projects.

REFERENCES

Alpiq. 2010. *Electra-Massa, Gebidem, Balades hydroélectriques.*
Boillat, J.-L., Dubois, J., De Cesare, G., Bollaert, E. 2000. *Sediment management examples in Swiss Alpine reservoirs.* Proceedings International Workshop and Symposium on Reservoir Sedimentation Management, Tokyo, Japan.

Boillat, J.-L., Dubois, J., Lazaro, P. 1996. *Eintrag und Austrag von Feststoffen im Spülkanal von Gebidem. Modellversuche und numerische Simulation.* Symposium an Verlandung von Stauseen und Stauhaltungen, Sedimentprobleme in Leitungen und Kanälen, 28./29. März 1996, ETH Zürich, Mitteilung Nr. 142 der Versuchsanstalt für Wasserbau, Hydrologie und Glaziologie (VAW), pp. 151–170.

Boillat, J.-L., Oehy, Ch., Schleiss, A. 2003. *Reservoir Sedimentation Management in Switzerland.* The 3rd World Water Forum, Challenges to the Sedimentation Management for Reservoir Sustainability, pp. 143–158.

Boillat, J.-L., Pougatsch, H. 2000. *State of the art of sediment management in Switzerland.* Proceedings International Workshop and Symposium on Reservoir Sedimentation Management, Tokyo, Japan, pp. 143–153.

Bosshard, T., Fischer, A., Kress, A., Kull, C., Liniger, M.A., Lustenberger, A. und Scherrer, S. 2001. *Swiss Climate Change Scenarios CH2011.*

EOS. 1999. *Chenal de la Massa à Bitsch, Remarques et réflexions.* From Bretz, N.-V. for Electra-Massa (unpublished).

EOS. 2002. Note interne no 2002/293, Chenal de la Massa à Bitsch, Conduite enterrée pour la dilution. (unpublished).

Istituto di Idraulica dell'Università di Padova. 1964. Essais sur modèles réduits des ouvrages de restitution de la centrale de Bitsch.

LCH-EPFL. 1994. *Etude expérimentale sur modèle hydraulique du chenal de chasse de la Massa à Bitsch.* Internal Report, Laboratory of Hydraulic Constructions, EPF-Lausanne, Switzerland.

LCH-EPFL. 2002. Gestion du transport solide lors de la purge annuelle du barrage de Gebidem. Modélisations physique et numérique des écoulements et du transport solide dans le chenal de chasse de la Massa à Bitsch. Internal Report n°8, Laboratory of Hydraulic Constructions, EPF-Lausanne, Switzerland.

Pougatsch et al. 2002. *Sécurité des ouvrages d'accumulation. Dam safety guidelines.* Swiss Federal Office for Water and Geology, Bern, Switzerland.

Schweizerische Gesellschaft für Hydrologie und Limnologie (SGHL), Hydrologische Kommission (CHy). 2011. *Auswirkungen der Klimaänderung auf die Wasserkraftnutzung—Synthesebericht.* Beiträge zur Hydrologie der Schweiz, Nr. 38, 28 S., Bern.

Reservoir Sedimentation – Schleiss et al. (Eds)
© *2014 Taylor & Francis Group, London, ISBN 978-1-138-02675-9*

Keyword index

Author index